APPETITE FOR LIFE

AN AUTOBIOGRAPHY

OTHER RECENT BOOKS

By Samuel A. Goldblith, Ph.D.

OF MICROBES AND MOLECULES
FOOD TECHNOLOGY, NUTRITION AND APPLIED BIOLOGY AT M.I.T.
1873-1988

PIONEERS IN FOOD SCIENCE™, VOL. 1
S.C. PRESCOTT, M.I.T. DEAN AND PIONEER FOOD TECHNOLOGIST

APPETITE FOR LIFE
AN AUTOBIOGRAPHY

by

SAMUEL A. GOLDBLITH, Ph.D.

FOOD & NUTRITION PRESS, INC.
TRUMBULL, CONNECTICUT 06611 USA

SAMUEL A. GOLDBLITH, Ph.D.

6 Meadowview Road
Melrose, Massachusetts 02176

Library of Congress Catalog Card Number: 96-061677
ISBN: 0-917678-38-9

Printed in the United States of America

DEDICATION

This book is dedicated to the family — past, the present and the future.

To my late, beloved wife of forty-nine plus years, Diana L. Goldblith, whose presence permeates and enlivens much of this book. For whatever success I may have had in my life, it is she who deserves most of the credit.

To my parents, Abraham and Fannie Goldblith and my grandparents, whom I never knew in person, but about whom I heard from my parents.

To Diana's parents, George and Ida Greenberg, who were like a second set of parents to me and to their progenitors.

To my children, Judith and Richard Clark, and Jonathan Goldblith, who have given me much pleasure and pride in their accomplishments and who have helped so much to lighten my load these past several years, so that I might enjoy as much of these "Golden Years" as possible.

To my grandchildren, Sarah Beth Clark and Rachel Deborah Clark. I am proud of you both and of your accomplishments. May your future be bright and full of Hashem's Manifold Blessings — especially good health, happiness and prosperity.

PREFACE

Why write an autobiographical memoir? Perhaps it is because if I don't do it, no one else will. Moreover, much of the material is in my head, or I am about the only one of my generation who knows where the material is. There is also the fact that for a number of years, some of my friends have been urging me to write of my experiences during World War II and of my relationship with a number of Japanese Nationals since the 1950's. Their reason basically was that I had lived through and participated in an important era in our national history and had survived an unusual set of events. My children were also interested in seeing such a memoir and adding to it some genealogical information about Diana's and my family as well.

Let me state at the outset that my knowledge of and participation in World War II was in a limited area of operations and only for a limited period. Moreover, for the bulk of that time, I was at the lowest officer rank. However, I was overseas when the war began. With respect to World War II, countless books have been written on various aspects of the subject — some by scholars and some by participants. In the case of the latter, I might add, the longer that one waits after the event, to write about the event and one's role therein, the more inaccurate the book probably becomes, either intentionally or unintentionally, due to the normal diminution of the human memory process. Moreover, intentional embellishment of what actually occurred is common, inasmuch as people often wish to be remembered in a certain way rather than by what actually occurred.

Thus far, I have published only three papers, of a technical nature, about some of my activities in World War II.[1] Those aspects of this present memoir that deal with World War II were written from memory, notes and a diary, jotted down within a year to two after the events while in Zentsuji POW Camp on Shikoku Island, Japan. This together with my "early years" plus the events that occurred post-Zentsuji until I arrived home after the war in

October, 1946 make up the first half of the book. The first chapter, "Early Years," was written over the past two years. The chapters dealing with Roku Roshi POW Camp, the return home and survival were written over the past six months, as I began to really shape the book.

The second half of the book dealing with a longer time span, from graduate student (1946-1949) to the present time, was written in various segments over the past year and one-half.

I hope that my accounts of the second part of the book express my own personal philosophy that times change and are altered by historical events. I do not believe that we can live with the immediate post-World War II mentality of hate and venom directed to our former enemies. This living in the past can only breed more venom and hate in the future. I do not think that we should ever forget the past history. Indeed, we must remember it and teach it accurately to our children. Revisionist histories are as abhorrent to me as is the "flag waving" of so many people and organizations. True patriotism is not to be equated with simply "waving the colors" and spreading hate about former enemies.

I belong to three organizations which are POW-oriented. One is the American Defenders of Bataan and Corregidor, the second is the Zentsujians, a small organization made up of former inmates of the POW Camp located at Zentsuji, Japan (located on the island of Shikoku). The third is the American Ex-Prisoners of War. These seem to have less "flag-waving" and less post-battle "rattle of musketry" than in the large massive organizations.

This autobiographical memoir presents a summary of the genealogy of Diana's and my families to the extent of information that I possess. Genealogy in our families was not as fashionable then as it is now. Moreover, my late beloved wife, Diana, and I were fortunate to have lived an interesting and busy life. We were fortunate in having met some remarkable personalities who have had unusual influences on me and mine. We also were fortunate in living in the age of jet travel so we saw a great deal. In writing this memoir I tried to highlight some of the interesting experiences that we both shared at home and abroad. The world is full of interesting people and Diana and I were fortunate to have met a few of them as I hope will be apparent herein. I hope that this document will encourage some in future generations of our family to keep updating it.

Isaac Bashevis Singer wrote in Love and Exile[2] :

> "As a believer in God and His Providence, I am sure that there is a full record of every person's life, its good and bad deeds, its mistakes and follies. In God's archives, in His Divine computer, nothing is ever lost,"

So it was with a feeling of humility that I undertook to write this autobiographical memoir. One cannot close a preface of an account of a full life such as Diana and I have led without offering praise to Almighty God for having given me the strength, the life and sustenance to have been able to reach this day.

Samuel A. Goldblith
Melrose, Massachusetts
June 18, 1996

REFERENCES

[1] Sanitation in Bataan and its Aftermath, S.A. Goldblith
The 803rd Eng. Bn. in the Defense of Bataan, S. A. Goldblith
(Both in the Military Engineer, circa, 1946)
Prisoner of War Camp Nutrition Study by Samuel A. Goldblith and Robert S. Harris
Monthly Research Report of the Office of Naval Research, March 1, 1949

[2] *Love and Exile*, Issac Bashevis Singer, Doubleday and Co., Inc., Garden City, N.Y., 362 pp.

ACKNOWLEDGMENTS

It is a pleasure to acknowledge the help and assistance that I have received from many and in particular from the following.

Mr. Carl A. Accardo, for the title of this volume. I asked him for a suggestion and received two or three. He and his son, Thomas A. Accardo, suggested the word "Appetite" tied in the central theme of some of my life's work with what these reminiscences are meant to do, cover the past but point to the future with zest for life.

Dr. Michael A. Kane, dear friend and healing physician who, as in the past two post-retirement volumes, edited, suggested and pushed me along to completion of the volume.

Ms. Margaret Duke, dear friend from the I.L.P., for volunteering to correct the manuscript on the computer — and doing it so well — something my failing eyesight could not do. My gratitude knows no bounds.

Mr. Murton Faibisch, Proprietor of The Camera Connection of Melrose, for his skill in converting large and small snapshots, color photos, old and new into a series of acceptable high-quality illustrations.

Mr. John J. O'Neil, Publisher of the Food and Nutrition Press for his editing skill and, together with his staff, especially Mrs. Maureen Yash, for the production of this volume.

My daughter Judith Clark and her husband Richard Clark for their help in recalling many of the family incidents recounted herein, as well as proper dating of many events and for their unfailing encouragement.

My son, Jonathan Goldblith, for his suggestions, proper dating of events, loan of pictures which I could not find — including many that I even took, and above all for his continuing encouragement to "the old man" when he needed it.

Ms. Rachel Clark, my younger granddaughter, for her help with all the multitudinous tasks that needed to be done as we neared the completion.

Ms. Donna Coveney/M.I.T. for the photograph of the M.I.T. Dome on the outside cover.

Mr. Thomas A. Accardo for the drawings of the maps of the Philippines used herein.

All the above goes to show that there is more to writing a book than the mere writing — "no man is an island unto himself." My humble thanks to all who contributed to its completion. Having said this, any and all errors of omission or commission are those of the author himself.

CONTENTS

CHAPTER 1

EARLY YEARS (1919-1941)

"Incipio": My Parents

I was born at the Lawrence General Hospital in Lawrence, Massachusetts on May 5, 1919 to Fannie (nee Rubin) Goldblith and Abraham Goldblith. My parents had been married sixteen years before I was born and had despaired of ever having any children. I am an only child. Thus, I was most welcome and, apparently the "apple of my parents' eye." The attending physician was Dr. Myer Schwartz, a general practitioner.

My father came to this country on April 15, 1911, aboard the S.S. Vaterland, landing at New York. Pa was born on August 20, 1878 at Ovelsk, Russia, which I believe was somewhere near Zhitomir. Pa had mentioned that it was a small village and large cities, not too far removed included Kremenetz, Kovno and Dubno. I cannot find Ovelsk on any map that I have. He had come over to this country under the name (on his ship's ticket) of "Schanie Gochman" for some reason which, if I ever knew, I have forgotten. I believe that, when they were married his name was "Abraham Feinblith." My mother's embroidered napkins, etc., which she had prepared for her trousseau, were done with the letter "F." I have Pa's draft card under the name of "Schanie Gochman, alias Abraham Goldblith." I believe, but do not know, that the Goldblith versus Feinblith was a product of Ellis Island Immigration Service. Pa registered for the draft on September 12, 1918 from 48 Manchester Street, Lawrence, Mass., Draft Board Division 3.

My father had worked in the so-called "sweat shops" in New York City after arriving in the USA to earn and save some money, so that my mother could come over to join him.

My mother arrived in New York aboard the same ship which left from Antwerp and arrived in New York on March 2, 1912. Ma's ticket is with the name "Feige Gochman of Wischniowic, Poland." From this, I deduce that Ovelsk must be

FIG.1.1. MY MATERNAL GRANDPARENTS, YECHIEL MECHIEL AND JUDIT CHANNAH RUBIN, TOGETHER WITH MY AUNT, MY MOTHER'S SISTER, CIRCA 1920'S

somewhere near the Polish-Russian border as it existed at that time.

They first settled in this country, I was told, in the lower East Side of New York City, where both my mother and father worked in one or another sweatshops. While there, my father contracted tuberculosis and spent over three years in the T.B. Sanitarium at Middleton, Mass. This was located very close to the present M.I.T. Bates Linear Accelerator Laboratory. The old T.B. Hospital structure was still standing some ten years ago when I visited the Laboratory.

FIG. 1.2. MY PARENTS, ABRAHAM AND FANNIE GOLDBLITH, TAKEN AT SANITARIUM AT MIDDLETON, MA WHERE PA HAD BEEN A T.B. PATIENT FOR SEVERAL YEARS BEFORE I WAS BORN, PICTURE DATE 1917

My parents moved to Lawrence sometime before the entry of the United States into World War I. I believe that they moved because my father was not well, having worked day and night to earn sufficient money to pay for my mother's ticket to this country. My mother had a brother, Abraham Rubin, who lived in Lawrence, as well as several "landsleite" from the "old country." This probably accounts for the move to Lawrence, plus the fact that the T.B. Sanitarium at Middleton was then among the best. Also my mother could live near her brother and friends whom she knew in Wischniowic. Somehow, over the years of hospitalization of my father, my parents and her brother became estranged. I never knew the reason, or if I did, I have forgotten it.

In those days, the only cure for tuberculosis was believed to be rest and fresh air and this treatment was what Pa received at the Middleton Sanitarium where he was a patient of a Dr. Sullivan's for about three and one-half years. This accounts for his not registering for the draft until September, 1918.

While in the sanitarium, Pa was very ill and for a while there was despair of his ever recovering. Under Jewish Law (Halacha), if something happened to Pa, my mother might be obliged to marry his unmarried brother(s) still in Russia. In order to protect my mother from this Halachic requirement, my father did an unusual thing. He gave my mother a Jewish Bill of Divorce (a "get," done properly by a Rabbi and a Sofer (scribe)). Thus if something happened to Pa, my mother would be free to do as she pleased regarding remarriage. She was no longer required to marry any of Pa's brothers.

Miraculously, Pa recovered from the dread disease and upon discharge from the sanitarium, he signed up for the Draft, but was never called. (I may add, parenthetically, that as a young man, he had served in the Russian Army Cavalry for several years before coming to the U.S., and was an excellent horseman. He had been brought up on a farm and knew horses, how to tell their ages, etc., as well as being an excellent rider).

During my father's lengthy hospitalization, my mother earned her living as a seamstress in Lawrence. I recall no further details. Upon discharge from the Middlesex Sanitarium, my parents received some help from Mr. and Mrs. Samuel Toabe of Lawrence with whom they were to become lifelong friends. I recall that when my parents celebrated their 25th wedding anniversary, (I was nine years of age at that time), two parties were held for them. One was at the Toabe's home on Olive Avenue, in the Tower Hill section of Lawrence and another was held at the home of Mr. and

Mrs. Joseph Issenberg in Methuen. I do not recall how the guests were split up. I only recall that I came with my parents to the Toabe's home and stayed with their two daughters. The elder, Sarah, later married Arthur Levine, a Professor at the University of Massachusetts at Amherst, and Ruth, the younger, later studied microbiology at the University of Indiana. The Samuel Toabes had a younger son, Jason (Jimmy), with whom I am still friendly and still see, together with his wife Florence, at the Ruth and Milton Issenberg home in Andover. Both are jewels of people.

The aid that my parents received from the Toabes, enabled them to set up a small grocery "mom and pop" type store and to purchase the flats around the store. They lived in the second-floor flat adjacent to the store. The flat was at Number 53 Manchester Street. The flat consisted of a large kitchen, a bedroom, (my parents room), as well as a bathroom off the kitchen. Across the hall was a large dining room and adjacent to that was a front porch which my father had enclosed and heated when I became eight or nine, as I recall it. My parents installed central oil heat in the late 1920's or early 1930's. They maintained a truly Jewish home, with a Mezuzah on every door, two sets of dishes — dairy and meat — and Sabbath Candles, etc. Rabbis and the Schochet frequently were guests at my parents' home, where they could eat with impunity.

Together with the income from the store and the nine or ten flats around the store, my parents made a living. It was hard work, but honest and they counted their blessings, gave to various charitable organizations, including the Congregation Anshei Sfard, "the Russische Schul," on Concord Street. They scrimped and saved, even in the height of the depression, so that later I could attend college. They were kind landlords and store proprietors, frequently extending credit to a number of customers. As I recall it, many of the tenants remained in the apartments, or flats, for extended periods of time — more than a decade.

My parents were great people, well-liked and respected. Among their friends were, in addition to the Samuel Toabes, the Samuel Issenbergs, the Joseph Issenbergs, a number of Jewish people that they had known in "the old country." They respected education and learning — both secular as well as religious. My father, being learned in the Torah and the Commentaries thereon, was friendly with Hazzan Zelig White, Rebbe Solomon Golubzik, my Hebrew School teacher. My father was close with an attorney in Lawrence, Atty. Maurice J. Dorgan, who came to the store or house frequently in the evenings and my mother fed him some Jewish delicacies. He was an alumnus of Philips Andover, Harvard College and Harvard Law and encouraged my parents to send me to college (not that they needed any encouragement in this regard). Later, the Dorgans attended the 50th Anniversary dinner party Diana and I had for my parents at the M.I.T. Faculty Club.

FIG. 1.3. AUTHOR ON PONY RIDE IN FRONT OF NEXT DOOR NEIGHBOR'S HOUSE ON MANCHESTER ST., LAWRENCE, CIRCA 1924

When I was young, circa five or six, I recall my father taking me to Boston by automobile (quite an expedition then), and as we passed M.I.T. he said: "Sammy, some day you will be going here to college. I hear it is the best in the world." I never applied to any other college other than M.I.T. I was lucky I was admitted and have never regretted studying at M.I.T.

Early Years

One of my earliest recollections as a little boy of two or three years of age, was attending the funeral of Samuel Issenberg's first wife, who had died shortly after a miscarriage. They had one son, Milton, with whom I have been very friendly ever since. Sam Issenberg remarried and had three sons with his second wife, Minnie. He founded the S. Issenberg Confectionery Co. which later grew into Lincoln Foods, but this is the story for Milton to tell.

Another early recollection I have is of my mother ironing at the kitchen table, while I was playing on the floor at her feet and she was singing a nursery rhyme to me, in Yiddish, "Auf dem Pripichik."

We spoke Yiddish at home. I went to school and learned English, above what I had learned on the street playing with the neighbors' children. I began school at five years of age, attending the Washington Street Primary School, only a block from our house. There, in the fourth grade, I had a teacher who had an important influence on my life, Miss Sarah Barlow. She tutored me in Latin and French, some evenings when I was attending high school. She was a great person and a fine teacher. She was very friendly with my parents.

When I began primary school, I also began studying at the Hebrew School, the Talmud Torah, on Concord Street which was located adjacent to the Anshei Sfard Synagogue. The primary courses were given by a Mr. Brown and after one or two years, we moved up to the larger room to be taught Chumash and the Tanach by Reb Solomon Golubzik (Rebbe Golubzik). Here, we not only learned to read and write Hebrew, but learned the important prayers as well as preparing for our Bar Mitzvahs. Rebbe Golubzik was a good teacher, but did not "spare the rod" in the form of a ruler. He was the grandfather of the three Averback boys, Joseph, Lewis and Philip, their mother Bella, being one of Reb Golubzik's two daughters.

FIG. 1.4. AUTHOR IN THIRD OR FOURTH GRADE AT WASHINGTON ST. PRIMARY SCHOOL, CIRCA 1928

My father used to say, with reference to Reb Golubzik's teaching, "one could always tell a Golubzik student by the fact that he never forgot his Torah Blessings and could chant them perfectly, many years later."

Rebbe Golubzik was also the Secretary of the Congregation and in those days, one bid for the Torah Honors. Since writing was forbidden on the Sabbath and Holidays, he had developed a unique system for keeping track of these individuals who were called up to the Torah by preparing in advance, pieces of paper, slit partway up longitudinal axis with various sums of money on the vertical axis and the various charities along the horizontal axis. He would bend over the tabs, in accordance with the pledges.

The synagogue and "Cheder" occupied a central role in the lives of the Jews and their youngsters in Lawrence, and throughout the Northeast. Thus, I dwell on it a bit. The Shachris (morning) before the Torah Reading, the Hazzan was Hazzan Zelig White, father of Samuel White (of Falk and White Wholesale Grocers), and grandfather of George White, and Edward White, with whom I am friendly to this very day, and Maurice, an Aero Engineer who graduated from M.I.T. about two years before me.

The Torah was read by Mr. S. Winer, a clothing merchant and a learned man in Hebrew.

My attendance at the Anshei Sfard Synagogue was mainly during the 1930's (other than the High Holydays). I was Bar Mitzvah at the Congregation Anshei Sfard in 1932, (Parsha "Emor," as I recall it). During the 1930's, the menace of Hitler was ever so vivid and ugly. I recall impassioned pleas made from the "Bimah" by leaders of the community, such as Abraham Rappoport, a local merchant, on Kol Nidre night in 1938 or 1939, pleas for money to save European brothers and sisters. Our parents and many other congregants had relatives in Eastern Europe and some in Germany. I shall always remember the impassioned appeals on Kol Nidre. With the Jewish belief that the saving of lives came first, even on Yom Kippur, secretaries were in the synagogue writing down the pledges — all rules for Yom Kippur were temporarily suspended for this fund raising appeal to send money abroad to help our Jewish brethren in Germany and Eastern Europe.

Among the other friends of my parents in addition to the Toabes and Issenbergs were the Hyman Hindmans. Mr. Hindman was a learned man and a shoe merchant on Essex Street. He was the father of two physicians, one of whom later delivered our daughter Judy. Both sons served in the Navy as physicians, but died early deaths due to kidney disease. Another friend of my parents was the Rev. Mr. Bulman, the Shochet, and Mrs. Bulman, who succeeded Reb Golubzik in teaching Hebrew School. Rev. Bulman, succeeded Mr. Winer as Torah Reader and also served as a certified Mohel, as well as Schochet. Mr. Bulman was from Poland and his wife was from France, a gorgeous lady and a lovely person. Mr. Bulman, took charge of the preparation of my father's body, after he passed away, for burial, so that he would be buried in accordance with proper orthodox tradition.

There was another orthodox synagogue in Lawrence with which my father became connected in the latter years of his life, before he came to live with us in 1964, the Congregation Anshei Shulum (The "Cross the Spicket" (river) Schul). When Pa could no longer walk up the hill to the Saturday and Holiday services he joined the Cong. Anshei Shulum as well, and chanted the Torah Readings every Saturday, Monday, and Thursday and on the High Holidays.

Other friends that my parents had were the Abraham Bakers and their daughter and husband, Rebecca and Charles W. Goldman, a Mr. and Mrs. Harry Fire and others. Abraham Baker served as the second president of the Congregation Anshei Sfard. Other friends included Miss Mary Toomey, an insurance agent in the Central Building where Maurice Dorgan had his office. She was Pa's insurance agent and later became mine — until the end of 1995 when she retired. She used to type notes from Pa and Ma to me in POW camps. He also had a friend in the Registry of Motor Vehicles in Lawrence where "Zadie" took Judy for her driving license test in 1964, while Diana and I were in Europe.

I attended the Alexander B. Bruce Elementary School from 5th to 8th grades. It was located on Ames Street in the Tower Hill area. My father felt that I would be among better people there than in the Tarbox school where I would normally have gone. I recall Pa giving me a ride quite often between the morning and afternoon sessions, so that I could eat lunch at home.

In addition to Hebrew School attendance, I also took violin lessons from a Mr. Herbert Middleton, who also had a band which marched in the Memorial Day parade each year, right by our house and store where we sold a lot of candy and ice cream that day. Sam Issenberg, after the parade in which he marched as a veteran of WWI, would come in with Milton and help sell.

Lawrence High School — (1932 - 1936)

I attended Lawrence High School for four years, graduating with the class of 1936 and received the Physics Medal. Between school, Hebrew School and violin lessons, I was kept rather busy. I do not recall when I quit Hebrew School, but it was probably after my Bar Mitzvah. Then I stopped taking violin lessons because High School required a great deal of home study. I enrolled in the College Preparatory Course. I took the usual program with language electives being two years each of Latin, French and German. I also took chemistry, physics, and mathematics through and including solid geometry and trigonometry. I graduated in June 1936 and was somewhere near the top part of the class, although I did not graduate with "distinction" or "great distinction." For some reason, I did receive the Riley Medal for excellence in Physics at graduation. I did join the Lawrence High School orchestra and as a result of this plus my mother and father's love of music, I became, in later years, a lover of classical music. My mother had a beautiful singing voice and her siblings all had played musical instruments or sang. It is from her that our son Jonathan probably inherited his beautiful robust tenor voice.

My father made sure that I had money so that once in a while I could attend the Red Sox or Braves games, to which I traveled from Lawrence by train and trolley. I had the privilege of watching such baseball luminaries as Babe Ruth, Lou Gehrig,

Lefty Grove, Lefty Gomez, Jimmy Foxx, Joe Cronin and Ted Williams, to name but a few.

The Y.M.H.A.

I have focused much of the foregoing around my Jewish upbringing and Jewish education. There was another aspect to Lawrence and Jewish life for young people and that was the establishment of a Young Mens' Hebrew Association (YMHA) and building therefor on Concord Street near Hampshire Street. This provided a focal point for Jewish young (and old) people to gather for indoor sports such as basketball, gymnastics, wrestling, etc. It had an athletic director, as well as an overall director, and an auditorium upstairs on the second floor. Lectures by various touring speakers and Yiddish plays by touring groups of actors were presented. The speakers that made a lasting impression on me were two. Dr. Abram L. Sachar, then the Hillel Director at Illinois or Indiana University, I forget which, and later to become the founding president of Brandies University, and the Reverend Dr. John Haynes Holmes, the famous minister and spellbinding orator of New York City who was a very close friend of Rabbi Stephen S. Wise.

A group of us in our very early teens formed a club known as the "A.A.F. Club" (Amici ad Finis). We had our own basketball team. Among the members were Milton Issenberg, Joseph and Lewis Averback, David Berger, Sidney ("Shyer") Rosenberg and a few others of our general age group. The "Y" was a good locus for us in a Jewish setting in terms of social activities, and many of us in the A.A.F. Club have remained friends to this day, although we may not see each other very often. I remained a member of the "Y" until I was called to active duty on April 24, 1941. But here, I am getting ahead of my story.

I had applied to only one college, M.I.T., and was accepted for admission in September 1936 with the Class of 1940. As I look back upon this, it was risky business applying to only one college, even then. But, it worked out all right.

FIG. 1.5. BASKETBALL TEAM OF A.A.F. CLUB OF THE LAWRENCE Y.M.H.A., 1934-1935
Author is to the right of center (front) and Shyer Rosenberg to the left of center; Milton Issenberg is at far right in second row.

Undergraduate Years at the Massachusetts Institute of Technology

At M.I.T., I had originally enrolled in Electrical Engineering, but soon found out for myself that my mathematics background at Lawrence High was insufficient for E.E. at M.I.T. After all, in my freshman year, we had many students at M.I.T. who had attended Phillips Andover, Exeter, Milton Academy, etc. who had already studied not only solid geometry and trigonometry, but also differential and integral calculus. Thus, they fared much better in math (many taking advanced standing courses), and also did much better in freshman Physics at M.I.T. which utilized calculus! Thus, I was at a disadvantage in Physics although doing passable work in calculus. So, I flunked the second semester of physics in my freshman year.

I recall that when I entered M.I.T., I also had the idea that I wanted to be a medical doctor. Did my parents put that into my head? I do not remember). Thus, I quickly switched my selection to a major to Biology and Public Health. The Department of Biology at M.I.T. was a relatively small department in terms of both faculty and students. It had three major areas of instruction: Biology, Biology and Public Health, and Industrial Biology. I was able to get to know all the faculty of the department and made friends with three or four very interesting faculty members even as an undergraduate. Later, in Graduate School at the Institute, I became even more friendly with these individuals, which I will discuss later on.

In June, 1940, I received a Bachelor of Science Degree (S.B.) in Biology and Public Health. The degree however, unlike most M.I.T. S.B. degrees at that time, was not a terminal degree. To become "someone" in this field, one needed a doctorate — and more particularly an M.D. degree. M.I.T. recognized this during World War II and did away with the degree programs in Public Health. I have discussed this in detail in the Biography on Dean Prescott published in 1993. My thesis for the S.B. Degree was in the then burgeoning field of nutrition under Assistant Professor Robert S. Harris. It was on the "Determination of Thiamine (Vitamin B_1) in the Urine of Pregnant Women." It was an exciting time in nutrition. The structure of the then known vitamins was being determined and as a result, rapid chemical methods of analysis were being developed. New vitamins were being discovered. This was the beginning of a new era in medicine — the Vitamin Era. In 1995, the vitamin business alone is a three billion dollar industry. It gave me a chance to participate in an exciting scientific adventure.

During my undergraduate years at M.I.T., I commuted from home in Lawrence via train (Boston and Maine Railroad), except for the first semester when I lived at the M.I.T. East Campus Dormitory, Goodale. Commuting was time-consuming, but one could study a bit on the train. I was a member of the "5:15 Club," the club of commuting students with headquarters in Walker Memorial. There I met Harold Hindman of Chelmsford, Massachusetts, who was a member of the Class of 1939 and whom I saw quite a bit of as we walked between M.I.T. and the North Station. Harold later became the co-founder of the Instron Engineering Corporation.

FIG. 1.6. AUTHOR IN M.I.T. UNDERGRADUATE CHEMISTRY LABORATORY, CIRCA 1937

During the summer between my sophomore and junior years, all majors in the Biology Department were required to take an intensive course in Organic Chemistry, summer of 1938. After the course was over, Harold asked me to join Ernest Kaswell

(M.I.T. 1939, later to become co-founder of Fabric Research Laboratories) and himself on a three-couple date. Kaswell's date was Miriam Hindman, Harold's sister; Harold's date, Evelyn, was from Haverhill. Evelyn later became Harold's wife. My date was to be a friend of Miriam's named Diana Greenberg, who later was to become my wife. Harold drove his mother's car (Kaswell having driven to the Hindman home in Chelmsford) and they picked up Diana in Lowell. They all picked me up in Lawrence and we drove to Haverhill to pick up Harold's date, Evelyn. All six of us (in the one car) then drove to the "Frolics" at Salisbury Beach, a night club with dance hall, where we had a grand time. As I recall it, none of us drank anything stronger than an orange juice. This took place on the eve of July 20, 1938 — my first date with my future wife.

I became friendly with Professor Harris, under whom I had done my S.B. thesis, and Prof. Murray P. Horwood, Professor of Bacteriology and Municipal Sanitation, under whom I did a long term paper on a Sanitary Survey of Lawrence, Massachusetts. Prof. Horwood wrote me a couple of long letters, while I was a POW in Japan which I received and still have. My father had visited him several times.

As an undergraduate, I was to meet Prof. Bernard Emerson Proctor, who was a protégé of Dean Prescott's. I noted that all of Prof. Proctor's students in Industrial Biology (as Food Technology was then called at M.I.T.) became employed immediately upon graduation. This was something I was to remember after World War II in my selection of graduate studies.

As an undergraduate, in my senior year while working on my thesis, I shared a laboratory with Edward Josephson who was working on the kinetics of enzymes at low temperatures under Dr. Irwin S. Sizer, Assistant Professor of Biochemistry. To finish our theses, we worked evenings and Diana, who was a freshman at the Massachusetts College of Pharmacy, used to come over and study in the laboratory and I would escort her back to the Franklin Square House on Washington Street, where she stayed five nights a week. Then, I would get on the El and take a late train back to Lawrence. Romance and science went together!

In order to earn a bit of extra change, I did some engineering drawings for another senior graduate student, John Cochrane Sluder, who was working under Prof. Proctor. Sluder was later to become Executive Vice President of The Nestle Company in the U.S.A. Thus, I not only got to know him, but Professor Proctor as well — the beginning of a long friendship. John Sluder passed away in the fall of 1995.

Reserve Officers' Training Corps at M.I.T. (ROTC)

During my first two years at M.I.T., I took the ROTC Courses which were compulsory for all male students for two years. At the end of my sophomore year, I signed up for the Advanced ROTC Program as a Reserve Officer Candidate in the Corps of Engineers. This ROTC Program required some extra courses plus one Summer Camp session of about six weeks, at Fort Belvoir, Virginia in the summer of 1939. My reason for taking the advanced program was that if our country went to war, which looked more and more likely, I would enter as an officer rather than an enlisted man. My selection of the Corps of Engineers was based on the fact that the Coast Artillery Corps, Signal Corps and Chemical Warfare Corps (which were available at M.I.T.) seemed to be further removed from my educational background than the Engineer Corps.

At Belvoir, we wore World War I uniforms, wrap leggings and all! We built bridges, learned to use road grading equipment, went on forced marches, etc. We lived in large tents, four men per tent, sleeping on cots under mosquito netting, which was good training for my future tour in the Philippines.

One of the officer candidates at Belvoir was a first cousin of Diana's who was a year behind me at M.I.T., Stanley Van Greenby (M.I.T. 1941). We traveled to Washington and Belvoir together. My father and Mr. and Mrs. Van Greenby saw us off at the South Station. Stanley's father, Frank Van Greenby, was a younger brother of Diana's father. He died an untimely death from cancer later that year. As I recall it, this was my first long train ride and first visit to Washington. Stanley and I spent at least one day each weekend exploring Washington, Mount Vernon and environs.

During our two years as Officer Candidates, we were paid at the rate of one dollar per day plus we were given an Officer's uniform, insignia, etc. One of the officers who was an instructor at the ROTC camp was Captain Frank E. Fries, CE. He was a West Pointer who was later to become the first Commanding Officer of the 803rd Engineering Battalion (Aviation), in which I served from July, 1941 through the World War II Campaign known as the "Defense of Philippines," entering into captivity together on the infamous Death March on the surrender of Bataan on April 9, 1942.

FIG. 1.7. AUTHOR AS UNDERGRADUATE R.O.T.C. CADET AT M.I.T., CIRCA 1938

I received a commission in the Army of the United States as a Second Lieutenant (Reserve) in the Corps of Engineers, a day or two before graduation.

1940-1941

What to do upon graduation? I was refused admission to Graduate School at M.I.T., probably because my grades in the first two years at M.I.T. were not too good. I received a Fellowship to study Nutrition at the University of Massachusetts under Dr. Helen S. Mitchell, but decided not to accept the offer.

I had worked, during my undergraduate years, part-time as a researcher or low-level clean-up man for Prof. Harris, being paid under the National Recovery Act (NRA) set up by President Roosevelt. Harris kept me on after graduation until about October, 1940 working with Mr. Malcolm Mosher, his Research Associate, doing odd semi-skilled laboratory jobs including the feeding of the laboratory rats. I found out I was allergic to rat dander and could not continue this part of the work.

First Real Job

Dr. Harris had been a consultant to Arthur D. Little, Inc., and one of their senior chemical staff, Dr. Stanley Cairncross, needed some assistance with a project he had for Bristol Myers, dealing with the over-the-counter product, "Sal Hepatica." I was hired by Dr. Cairncross to do an extensive literature survey on information on laxatives in the G.I. tract. I recall that Cairncross who had a Ph.D. from Columbia, as well as having worked in pharmacy, started me off on Goodman and Gilman, a textbook on Pharmacology, still in use today. Through this employment, I met with Mr. Raymond Stevens, Vice President of Arthur D. Little (MIT '17), Loren (Johnnie) B. Sjostrom, who later became head of the Food and Flavor Group at Arthur D. Little, and a number of other prominent staff members at Little. We prepared literature searches on foods and pharmaceuticals as a basis for new product development. In those days, computerized literature searches were a long way off. We did everything in pen and ink on 5 × 8 cards with abstracts thereon, shuffled the cards according to an outline of the report we had prepared and laboriously wrote and rewrote the reports.

I was earning about thirty-five dollars a week and although this was at that time, a respectable sum, I could see that I would never get too far in any scientific field without an advanced degree. But that would, as circumstances developed, have to wait.

Diana Lillian Greenberg, My Future Wife

From the summer of 1938 through 1939, 1940 and 1941, I was dating Diana. As mentioned earlier, I had met her on a blind date after the Course in Organic Chemistry was completed, on July 20, 1938. Diana Lillian Greenberg and her parents lived on Gibson Street, later moving to 24 Holden Street, Lowell, Massachusetts. Diana was a petite, blonde, beautiful girl, about two years younger than I. She was vivacious, extremely bright and fun to be with. She was much brighter than I and I learned much from her. Whatever success I may have had in life, much should be attributed to her.

Following her graduation from Lowell High School at a young age of 16 (having skipped two grades), in 1937, she took two years of post-graduate studies at Lowell High School and then entered The Massachusetts College of Pharmacy on Longwood Ave. She resided, five nights per week, at the Franklin Square House which was then situated at the corner of Washington Street not too far from the Dover St. Elevated Station. She went home by train for the weekends. She would study in the laboratory where I was doing my thesis work some evenings and I would take her "home" via the subway and then proceed to Lawrence by subway and train. We would sometimes have a fish dinner in Park Square where there was a famous fish restaurant (Pierroni's) where one could get a very fresh, complete fish meal for ninety-nine cents.

I visited Diana at her home for a Friday night dinner fairly often. We would go out on Saturday evenings. I would take the bus to Lowell and then sometimes we could borrow her father's car.

FIG. 1.8. DIANA LILLIAN GREENBERG, FUTURE WIFE, DRESSED FOR A FORMAL FRATERNITY PARTY AT LOWELL TEXTILE INST. (NOW UNIV. OF LOWELL), 1937-1938
Author carried this picture in his wallet throughout World War II.

I was working at Arthur D. Little through the better part of the period, August, 1940-April, 1941 and while I was not earning a tremendous amount of money, it was enough to support a wife in those days. After much discussion with parents, etc., we made plans to get married on April 27. 1941. The discussions had to do mainly with age (Diana was nineteen when we became engaged). I was an only child and the "apple of my parents' eyes," and they felt that I was too young to get married. We persisted and won out.

Diana's Parents and Their Family

Diana came from a lovely family. Her grandfather, Ezriel Greenberg, was a builder and owner of the Lowell Lumber Company. Her grandmother, on her father's side was the sister of Grand Rabbi Z.Y. Friederman, who married Diana's parents and Bar Mitvahid her only sibling, her brother Irving. Diana's father, George Greenberg was the first Jewish baby born in Lowell. Her grandfather, Ezriel Greenberg, built many two-family houses in the Westford Street areas of the city.

FIG. 1.9. GEORGE AND IDA GREENBERG, DIANA'S PARENTS, TOGETHER WITH HER BROTHER IRVING AND DIANA, CIRCA 1924

Diana's mother, Ida Frances Greenberg, was born in Sheffield England and came to this country as a babe in arms. They moved to Lowell when she was very young and Diana's mother attended Lowell Public Schools through graduation from Lowell High School. Women High School Graduates were not that common in Diana's mother's time, circa 1905-1910. Her father, David Baker, had been a paper hanger and painter in England. The family originally had come from Lithuania. In Lowell Ida's father, David Baker, opened a "Raquet Store" as it was known in those days. They sold dishes, pots and pans, and other housewares.

Diana came from a large extended family. During my time, the following are of note for the succeeding generations. Diana's father's younger brothers, included Frank (who died in 1938 or thereabouts) who was married to Aunt Etta. They sired cousins Donald and Stanley Van Greenby, both of whom went to M.I.T. Uncle Max Greenberg, who lived until the late 1980's, a veteran of WWI Balloon Corps, and an ex local actor, featherweight pugilist who fought in Boston under the name of "Kid Green," a former automobile car dealer who sold "Hupmobiles," and latterly ran a home "bookie" on the horses, until he "retired" circa 1955. He was a great guy and I got to know him and his wife, Marion, very well. Then there was Uncle Sidney Greenberg who died in the 1950's or early 1960's. He was the proprietor of the Lowell Lumber Company. Diana's father also had a sister, Lillian, who died at an early age. She married a manufacturing jeweler, David Goldwasser who made our wedding and Diana's engagement rings. David and Lillian Goldwasser had a daughter and two sons. The daughter, Diana's cousin, Gail Goldwasser, was like an elder sister to Diana as she was growing up. Gail had two brothers, Solly, who died while a student at Harvard Medical School and Milton, who also went to Harvard and became the Classified Section editor of the Los Angeles Times newspaper until he retired.

On Diana's mother's side, there were her two brothers, Uncle Solomon Baker (who married Aunt Sophie (nee Friedberg) who died of cancer in 1947). Solomon ran the family Raquet Store until his wife Sophie decided that he should become a lawyer and that is exactly what he did. He graduated from Boston University School of Law, passed the Bar Examinations and practiced law in Lowell for many years. They had a son, David A. Baker, who also went to B.U. Law School, after finishing his BA degree at Brown, and now lives in the Chicago area. Diana's mother had a younger brother, S. Joseph Baker, who became chief first-aid man of the Billerica Car Shops of the B & M Railroad. Joe was nicknamed "Iodine Joe" by his patients. He was a great guy also and baby sat for us for a few years after World War II.

Both of us came from very Jewish homes. Friday night meals at Diana's parents' home was a special delight with a meal that began, as I recall it, with chicken soup, then chopped liver, followed by roast chicken with potatoes, homemade baked beans, a cooked vegetable and a salad, and ended with homemade lemon meringue pie or sometimes, apple pie. What a meal!

Marriage of Diana L. Greenberg and Samuel A. Goldblith

A date of April 27, 1941 had been set for our wedding. Because of Jewish Holidays, etc., there was but a limited window of possible dates on which we might be married. Lo and behold, earlier in April, I believe it was then, or it could have been in March, I received "greetings from Uncle Sam" ordering me to Active Duty as a Second Lieutenant in the Corps of Engineers, Reserve, and ordered to report to the Commanding Officer of the First Engineer Battalion of the First Division ("The Big Red One") for a one year tour of duty beginning April 24, 1941. The Division Commander was General Terry Allen, and the Assistant Division Commander, who remained such until his death on the beaches of Normandy of a heart attack, was Brigadier General Theodore Roosevelt, Jr., son of the thirty-second President of the United States whom I met on maneuvers later in May 1941.

The Executive Officer of the First Engineer Battalion, to whom I reported on April 24, 1941 was Captain Charles T. Tench.

At the end of World War II, it was the same C.T. Tench, now a full Colonel, who was the first officer to set foot on Japanese soil on August 28, 1945, where he was greeted by Lieutenant General Arisue prior to General MacArthur's arrival. Tench's mission was to clear the way for MacArthur's arrival and to have everything set up in advance. At that time, I was still in a POW Camp at Roku Roshi, Japan, perhaps one hundred or two-hundred air miles away — it's a small world.

Having explained the situation, regarding the wedding, caterer, etc., in advance, to Captain Tench, he felt that the simplest thing to do was to have me report on time for active duty on April 24th. I immediately was given a four- or five-day leave of absence, as I recall. I left Fort Devens on April 26th, to get married on the 27th.

Diana came to the wedding in Uncle Sidney Greenberg's car — he always had the biggest car in the family, I am told. April 27th was a rainy day but the skies cleared just in time for Diana's arrival at Temple Beth El.

Diana looked lovely in her wedding gown. I wore tails and a top hat. The wedding was conducted by the Rabbi of Temple Beth El of Lowell, Rabbi Joseph Warren. Diana's father and Uncle Sol, helped found and build Temple Beth El and Uncle Sol had been one of the early presidents. Rabbi Warren had taught Diana Hebrew and other Jewish studies leading to her confirmation by him. Rabbi Warren was at his best, having known Diana for many years and her parents were friends of the Warrens, as well as being pillars of the community. Moreover, my father had gotten to know Rabbi Warren and thought him very well-educated and learned in Torah and Halacha. Rabbi and Mrs. Warren's son, David, went to M.I.T. and later became a patent lawyer in the New York area. I recall seeing David Warren in the 1960's or 1970's when he brought his son to look at M.I.T. for possible admission. I don't remember whether he enrolled or not. The kosher-catered full meal was superb, with champagne, wedding cake, dance band, etc. Everyone there, who heretofore might have been ignorant of the fact, learned that I was not a good dancer!

Miriam Hindman, who introduced us on July 20, 1938, was the Maid of Honor for Diana and Sidney ("Shyer") Rosenberg, one of my classmates at Lawrence High School and at "Cheder," was my best man. Shyer Rosenberg's father, David Rosenberg, had been a long time President of the Congregation Anshei Sfard on Concord St. and during that period, the Rosenbergs lived on Concord Street directly across from the "Schul." I recall vividly that there was a fire in the Synagogue and the Ark holding the Holy Torah Scrolls was beginning to burn. David Rosenberg ran into the synagogue and passed the Torahs to his boys, thereby saving the most valuable possessions of the synagogue. David Rosenberg was a brave man and the synagogue and its Torahs meant much to him.

The ushers were Dr. Edward Josephson, with whom I shared the laboratory for our respective theses — he for his doctorate and I for my S.B. degree, and Harold Hindman, Miriam's brother and through whom I met Diana and with whom we shared our first date. Harold took pictures at the wedding but for some reason, they never came out. All we had was a newspaper clipping from the Lowell Sun newspaper with Diana's picture therein.

After the wedding, Shyer drove Diana and me to Prospect Street in South Acton, where we had rented an apartment on the lower floor of a two family house, so that I could commute to Fort Devens. The next evening, Monday, the 28th of April, my father, together with Diana's father drove up from Lawrence/Lowell to deliver to us my parent's wedding gift to us, a secondhand DeSoto Coupe. Diana and I then drove both fathers back to her parents' home in Lowell and we returned to South Acton.

I commuted to Fort Devens almost every day from South Acton, except for the nights when I was "Officer of the Day." We had a short honeymoon. Poor Diana had to set up housekeeping immediately. She was a great cook and I began to gain weight reaching almost 180 pounds, never again to reach that weight until I was in my sixties!

First Engineer Battalion of the First Division

I do not remember much about the battalion since I was an officer therein for such a short period. We went on maneuvers, built bridges, etc. The training was typical of the pre-war period. Among a few of the "yellowed" papers that I have of that period is one detailing me as Assistant Judge Advocate for a Special Court Martial beginning on May 21, 1941. The Court (judges) were a group of seven officers headed by Captain Tench, the 1st Eng. Bn. Executive Officer.

Among the new reserve officers assigned to the First Engineer Battalion included several with whom I would serve later during World War II in the Philippines. These included:

Captain Herbert V. Ingersoll
First Lieut. Edmond P. Zbikowski
Second Lieut. David B. Bartlett
Second Lieut. Robert Montgomery

Ingersoll became Company Commander of Co. B. of the 803rd. Eng. Bn. He was a great man who died during the latter days of imprisonment, I believe aboard a Japanese ship being transported to Japan from Mindanao. Diana became friendly with his wife, Geraldine, and I saw her after the war. She was a lovely lady.

Lieut. Zbikowski was killed during a bombing attack on Corregidor where he was commanding Co. A. I do not recall what happened to Bartlett or Montgomery.

Transferred to 803rd Engr. Bn (AVN) (SEP)

We were just beginning to learn the duties of an Engineer Corps Officer when we suddenly, on June 21, 1941, received orders to report to Westover Field in Chicopee, Massachusetts (near Springfield) (the four officers above and me) to help activate a new engineer battalion which the army was creating, the 803rd Engineer Battalion (Aviation) (Separate). The Army was creating two such battalions designed to build and repair airdromes.

So, Diana and I packed our clothes, dishes, pots and pans, etc., said good-bye to our parents and drove to Westover Field. When we saw the sign on Route 20, I believe it was, "Boston, 90 miles," we both shed some tears. We rented a small apartment in Chicopee. All I recall about it is that it had a Murphy bed. I reported for duty at Westover Field as prescribed.

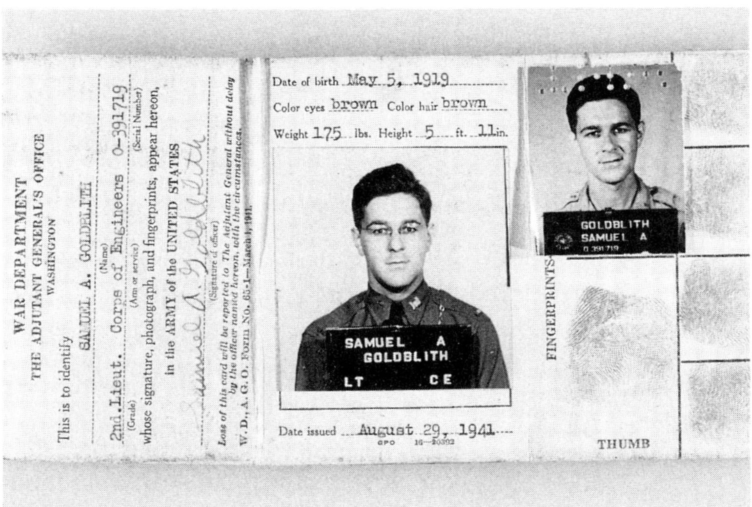

FIG. 1.10. AUTHOR'S U.S. ARMY I.D. CARD, AUGUST 1941

The 803rd Engineer Battalion (Aviation) (Separate) was activated on July 7, 1941 at Westover Field, Chicopee, Massachusetts with a cadre of 3 Regular Army officers, about 15 Reserve Officers, 40 enlisted men from Fort Riley, Kansas and about 26 enlisted men from an Engineers' Post in South Carolina. Most of the troops were draft inductees who came to Westover from the Training Centers of Fort Belvoir and Fort Leonard Wood.

My initiation period in the army was over, and the real work was about to begin. Poor Diana, she bore the brunt of finding new places to shop, meeting new faces, etc. She was inherently a shy person, so it was not easy for her. Our parents visited us on a couple of Sundays, while we were in Chicopee. Personally, we were very happy with our married life. But the first taste of the army aspect of get up and move, etc. was not to our liking. But after all, I was only in for one year from 24 April, 1941, according to my orders! Wow, were we to be surprised.

CHAPTER 2

THE 803RD ENGINEER BATTALION (AVIATION) (SEPARATE) WESTOVER FIELD, MA TO PHILIPPINES (JULY 1941 TO OCTOBER 1941)

In addition to the officers noted in the last chapter to be transferred, there was Second Lieut. Robert A. Montgomery. The officers from the First Engineers who were transferred were stationed as follows.

Capt. Ingersoll to Hq. Co.
1st Lt. Zbikowski to Co. A.
2nd Lt. Bartlett to Hq. Co.
2nd Lt. Goldblith to Hq. Co.
2nd Lt. Montgomery to Co. A.

The bulk of the battalion enlisted men were selective service draftees. As mentioned in the previous chapter, a number of non-coms arrived shortly after we did to act as cadre to help set up the battalion and to assist in training the draftees. These included 9 sergeants, 3 corporals, 9 P.F.C.'s (Private First Class) from the Second Cavalry Division at Fort Riley, Kansas plus a further complement of one sergeant, 2 corporals, 3 P.F.C.'s and 13 privates from Fort Jackson, S.C.

The battalion was to be commanded by Frank E. Fries, Captain, C.E. (who was about to be promoted to Major). The Executive Officer was to be Captain Clarence E. Bidgood, C.E., and 1st Lt. James D. Richardson, C.E., who was to serve as adjutant. All three were West Pointers and Regular Army Officers. Training began on June 14th, 1941.

Additional officers and enlisted men were added so that we had a complement of 3 captains (2 regular army), 8 first lieutenant (one regular army), and 7 second lieutenants (one-regular army,

ROTC Commissioned). In addition, we had 3 medical officers, one captain (Sidney Vernon), two first lieutenants, (Herbert Coone, Medical Corps and a dentist, Dr. Novack).

Training began almost immediately. We were equipped, in addition to pistols and Springfield rifles, with heavy road grading equipment including, cranes, bulldozers, road graders, etc. We trained on the heavy equipment, officers as well as enlisted men, by building a new runway at Westover Field. I recall learning how to operate a road grader and almost being hit by the head of a tall crane, which suddenly broke loose. I leaped away at shouts of "watch out" and only got nicked on my right ear and I had to have it plastered against my head so that it would heal and thereby reattach itself. I had a lucky break! Thus, we spent the hot summer at Westover Field, with occasional night duty as O.D. (Officer of the Day).

Diana and I became friendly with another second lieutenant and his wife, James L. Leggett, Jr. and Betty. Betty and Diana got to know many of the officers' wives, all of whom remained in touch with each other later when we went overseas, et sequel. As mentioned earlier, Diana and I saw our parents when we could for part of the weekend, inasmuch as I worked 5 1/2 days per week at the 803rd Engineer Bn. We decided to purchase a few pieces of our own furniture, some time in July or August just before a decision was made that the 803rd Eng. Bn., being more combat ready than the sister aviation Engineering Battalion that had been formed at the same time, was to be sent to aid General Douglas MacArthur in his building up the defense of the Philippines. So we canceled the furniture order. All our possessions were being shipped by the Army to Diana's parents' home at 24 Holden St., Lowell, MA. Since Diana had not learned how to drive at that time, my father came to pick up the car and say goodbye to me. It was a heart-wrenching leave-taking. I accompanied Diana to the railroad station at Springfield, MA, where she took the train to Lowell — another sad event. How long would we be separated? Little did we know what would transpire in the interval, nor really how long it would be before we saw each other again.

Meanwhile, Diana was pregnant. Our thinking was that inasmuch as we both wanted a child, we might just as well let the government pay for the birth expenses. She had been examined at the Westover Field Hospital and found to be just fine.

Transfer to San Francisco for Embarkation to the Philippines

On September 20, 1941, the battalion (Hq. Co., Co. A. and Co. B.) left by a train that consisted of 17 Pullman Cars from Springfield, en route to San Francisco, California where we were "to report to the Port Commander, San Francisco, California, Port of Embarkation for further instructions." The last paragraph of the orders detailed yours truly as a "Class A" Agent Finance Officer to purchase ice enroute, not to exceed thirty dollars ($30.00) and to account for this on arrival at San Francisco when I would be relieved.

The officers had lower berths in the Pullman Cars. It was a long train of Pullman Cars, together with several dining cars. Rosh Hashana, was September 22, 1941 which I "celebrated" enroute, on the special troop train, to California. We made a four hour stopover at Kansas City, where some of us headed for a good steak restaurant and 125 soldiers showed up at the dispensary car, once the train got rolling[1], for prophylaxis! Four hours was long enough a stopover. We had a short stopover at Albuquerque, NM. It was a long and boring trip.

There is nothing like a train ride to California to see the countryside. I really did not appreciate this on that occasion, but Diana and I took a trip by train cross-country in the late 1950's when I had some consulting in the Bay Area. By then the cars were air-conditioned and we had a bedroom. I recall that we came back to Chicago via the Feather River Canyon. It was gorgeous. There's a proper time for everything.

The Bay Area and Warm Personal Reception

On arrival in the Bay Area, at Oakland, as I recall it, we were transferred and transported by bus and boat to Fort MacDowell on Angel Island. We were in double-decker bunks, as I recall it. The officers were fairly free to come and go to San Francisco via ferry with proper signing-in and -out procedures, of course. The ferry made a stop at Alcatraz Island, which contained that infamous Federal Prison for the worst of the federal prisoners. It has since been closed down.

My recollection is we had fairly good freedom of access to San Francisco. Our main duties at Fort MacDowell were to be sure that we had all our equipment and personal effects, receive our immunization shots and had a physical. As I recall it, we received no orientation at all. During the week or ten days that we were on Fort MacDowell, the Holiest of Days of the year in the Jewish Calendar occurred, Yom Kippur, October 1, 1941. I wanted to go to a synagogue and there were no Jewish Chaplains on Fort MacDowell.

I knew that one of our enlisted men, a draftee, Joseph Fox from the Bronx, with whom I became acquainted, was Jewish and religious. He and his brother had been furriers in New York and they had family friends in San Francisco, who was also a furrier, Mr. and Mrs. Stgmund Haber. The Habers had a large fur shoppe on the mezzanine floor on Geary Street in San Francisco, opposite the then extant City of Paris Department Store. Joe was invited to the Habers for the pre-Kol Nidre meal, to services and "to break the fast" with them. Joe was certain that I would be most welcome at the Habers and indeed I was. They were lovely, hospitable people. I do not remember much about the service except that it was in an orthodox "schul" and I recall the meals to be superb and "hamisch." It was to be my last Kol Nidre for five years!

I jump ahead of my story, but I should point out that on the way back from the Far East after the war was over, I dropped in on the Habers at their Fur Shoppe, while I was a "patient" at Lederman General Hospital being checked over again, in October,

1946. The Habers were very pleasantly surprised, astounded in fact, because, Joe Fox had returned a week or two earlier and he had heard that I had not made it. The Habers and I had a joyful reunion and, of course, some delightful meals together.

I wanted Diana to meet them and we visited them several times after the war, in the 1950's and 1960's. They were lovely people and most gracious to a somewhat "homesick" young man of 22 who was then farther from home than he had ever been before in his life. I could not help but admire my own parents, traveling to a "foreign strand" in 1911 and 1912, to make a new start in life in America.

Once I knew I was going overseas to the Philippines, Diana and I talked quite a bit and consoled ourselves by feeling lucky that I was going to the Philippines rather than to England, where the Continent was in full blaze and England was being bombed daily. How wrong we were was to be proven shortly thereafter by the events of the bombing of Pearl Harbor and Clark Field.

Crossing the Pacific to Hawaii, Guam and the Philippines

Sometime during the first week of October, 1941, our 803rd Eng. Bn., was taken aboard the U.S.S. Tasker H. Bliss, a troop transport that formerly had been the S.S. President Cleveland of the President Lines. As I recall it, this ship was loaded with 1,500 enlisted men and 127 officers. Our equipment was jammed in the hold of the ship. It traveled at 15 knots per hour and I shared a stateroom with two other second lieutenants, which in peacetime was a stateroom meant for two people.

The trip was boring. We exercised by walking around and around the ship, did some calisthenics, read and ate. The food was pretty good, as I recall it. We passed under the Golden Gate Bridge just at dusk and sailed into the Pacific which at that particular time chose to be not so peaceful. A storm was raging. The Tasker H. Bliss rolled, heaved and yawed and many of the enlisted men below became sick. They had just consumed their evening meal of frankforts and sauerkraut. The lavoratories were overflowing. I went down to look after my men and talked many of them into coming on deck and getting some fresh air, which helped quite a bit. I was not feeling too "ship-shape" myself after the trip down below, but the fresh air helped.

We arrived in Honolulu and were met by a brass band. We remained there for about 48 hours to refuel, take aboard fresh provisions and water. Officers and enlisted men were permitted shore leave. I had a one-track mind. This was Hawaii and I could get some fresh pineapple juice, not the canned stuff we got on the East Coast. I and a fellow officer went to a juice bar and ordered fresh pineapple juice. Were we surprised — canned pineapple juice arrived. What a disappointment!

We made an additional stop for about five hours at Guam to take on fresh water. I recall going ashore, by lighter, for two hours or thereabouts at Guam. But there was little or nothing to see at Agana.

I should mention that once we left Honolulu, we were in a convoy with one other troop ship plus a cruiser escort. We were under total blackout conditions, port holes covered, no cigarette smoking on deck, etc.

Arrival at the Philippines

We passed Corregidor, as we entered Manila Harbor on October 23, 1941 and docked at Pier 7. Again a brass band greeted us. It had taken almost a month to cross the Pacific!

We were immediately transported to Fort Stotsenberg in "Pampanga" busses (extra wide busses owned and run by the Pampanga Bus Co.) and to be used, oh too soon, during the war for Philippine Army troop transport. We arrived late at night on October 23-24, 1941 from Manila, and were taken directly to Fort Stotsenburg.

We were housed, initially in officer's tents, two lieutenants per tent, along the main road into Fort Stotsenberg. Across the road from us was a troop of the 192nd Tank Battalion (National Guard) with eight light tanks, equipped with 37 mm cannons.

Fort Stotsenburg was the home of the 26th Cavalry (Philippine Scout) Regiment which was a famous horse cavalry comprised of Philippine enlisted men and non-commissioned officers with U.S. Army Cavalry officers. It was an old organization with some of the non-coms being second generation. They were well-trained, brave and capable men and officers. The officers of the 26th Cavalry had homes along two sides of the parade grounds in the center of Stotsenburg.

Our enlisted men moved shortly from tents in Stotsenburg to Clark Field Air Base into Nipa Barracks which were light and airy. The men (and the officers) slept under mosquito netting. Clark Field was located just across the main highway, Route 3, which ran from Manila-San Fernando northward to Damortis on Lingayen Gulf. Clark Field was one of the main airfields in Luzon housing some pursuit squadrons and about 30 plus of the newly arrived B-17's — the bulk of MacArthur's heavy bombers (35 B-17's). The junior officers were housed, beginning some time in November, in Nipa Barracks, BOQ (Bachelor Officer Quarters), one officer per "room," on cots with mosquito netting, and a single light bulb hanging down by a wire in each room.

The Battalion acquired the 809th Engineer Company (Aviation) (Separate) which became Company C of the 803rd. The 809th had been and would remain at Nichols Field just outside Manila, where, I believe the Manila International Airport is now located.

It was our primary responsibility to keep the runways, at Clark Field, in good operating condition at all times and to build on to the existing runways. For this purpose, Headquarters Company, of which I was a member, remained at Clark Field. We were kept busy maintaining the runways, training the enlisted men, many of whom were raw recruits and draftees, and some of whom could barely read or write!

FIG 2.1. NIPA BARRACKS FOR HQ. CO., 803RD ENG. BN., CLARK FIELD, LUZON, P.I., NOVEMBER 1941

Company A under First Lieutenant Zbikowski, with Dr. Coone, as Medical Officer, was sent to O'Donnell, Capas, Tarlac, later to be the site of the infamous O'Donnell POW Camp, where most of the survivors of the Bataan Death March (I included) were sent. Their mission was to build a landing strip suitable for heavy bombers (B-17's). O'Donnell was about thirty miles from Clark Field.

Company B, under Captain Herbert V. Ingersoll, was sent to the Sugar Central area of Del Carmen to construct a complete airdrome there, complete even with underground facilities.

During early to late November, I recall making a weekend trip or two to Manila and on one of these, hitched a ride in the nose of an old B-10 bomber. For the return trip, as well as on other trips, I took the train from Angeles to Manila. I went with some fellow officers from the 803rd, including Lieut. Mowack, the Battalion dentist, and Lieut. Radcliffe from Headquarters Company. I recall that we stayed at the Luneta Hotel on Luneta Park and of visiting my classmate, Oscar B. Mapua on Taft Avenue, Ext. Pasay, at the family home. We have remained in contact ever since. Oscar's father founded the Mapua Institute of Technology, "the MIT of the Philippines," and Oscar is now the President.

I wrote frequently about our going's-on and mailed weekly to reach the Pan American, weekly Clipper mail, which then cost 50 cents, U.S., per one-half ounce! I still have most of the letters Diana sent me. I also recall that evenings during November and the first week of December, we went to the Officers' Club for a night snack of hamburgers, malted milks, ice cream, etc. I had a white uniform made for a relatively few dollars by an itinerant tailor who arrived to take measurements that day after we arrived at Clark Field. Some of the enlisted men "shacked up" with young Filipino girls. It did not take them long to become "comfortable."

FIG. 2.2. AUTHOR AT LUNETA HOTEL, MANILA, NOVEMBER 1941

At Clark Field, we received intelligence, during the last week of November, that the Japanese envoy talks (Special Envoy, Saboro Kurusu) in Washington were breaking down and we had better be prepared for war with Japan. This we learned at an "all officer" call of officers from all units stationed at Clark Field and at Stotsenburg. I recall, that as a result we dug with our backhoes and other trench-digging equipment, "V" shaped trenches near the barracks at Clark Field, revetting them with woven bamboo and mounted a WWI .30-caliber machine gun at one end of each trench.

We also aided in the construction of revetments for our P-40 pursuit planes. These revetments were made up of gasoline drums filled with sand and stacked two high. In addition, of course, maintenance of the existing runways at Clark Field was carried on, as well as a new heavy bomber runway begun.

REFERENCE

[1] Coone, Herbert W., 1992, *The Sequential Soldier,* Gateway Press, Inc., Baltimore MD, 262 pp.

CHAPTER 3

WORLD WAR II IN THE PHILIPPINES (DECEMBER 7-8, 1941 – APRIL 9, 1942)

First Three Weeks of the War

At 12:30 PM on December 8, 1941 (Philippine time and date), we were attacked at Clark Field by 54 two-engine bombers, followed by Zero fighters. This was right in the middle of the lunch hour. I was in the BOQ, having had lunch with my men, and heard screams of the wounded, the loud explosions of bombs, strafing from the "Zeros," and saw the Japanese planes with the big red circles on the under side of the wings. There was a black pall of smoke over the airfield with flames shooting out of it. We jumped into the nearest V shaped trench and I manned the .30 caliber machine gun (WWI vintage) at the end of our trench. I was trying to get a lead on the planes, but unfortunately did not get a single hit. The Zeros came over us again and again, strafing us and all around us. I kept on firing until the gun jammed. I took out my .45 pistol, in a fit of desperation, and fired away with, of course, no luck at all. The Japanese fighter planes came over us only 200 to 300 feet off the ground. They were so close, we could see the pilots' faces and features. Our own casualties were light — two officers slightly wounded. None of the men was harmed. The slit trenches (plus luck) worked. Immediately following the attack, we sent out work crews under the command of an officer, salvaging trucks and equipment and repairing the runways which were badly cratered. By dusk, a landing strip for our planes was ready for landing and a few takeoffs. The process of being bombed and repairing runways continued until Clark Field was evacuated. On the Saturday morning following the initial attack, we suffered eight raids on Clark Field on that day alone. In the meantime, our heavy trucks and drivers were very busy hauling 8-inch guns, which were stored all over Luzon, to the Bataan peninsula, where temporarily, at least, they would be out of the enemy's reach.

The bombers and fighters created a great deal of carnage on the airfield, and destroyed most of the planes on the ground, which were the P-40 fighters and the B-17 bombers. Unfortunately the revetments we had labored to build had not been of much help. The Japanese destruction of the heavy bombers had been made easier by the fact that the B-17's had been out all morning and had just returned a short while earlier for lunch. They had been in the air to keep them out of harm's way while General Brereton tried to get MacArthur's permission to bomb Formosa. The B-17's were lined up to permit them to refuel rapidly and to load them with bombs — after lunch. Moreover, they were lined up wingtip to wingtip to prevent sabotage. They were like sitting ducks for the Zeros.

Late that first night of the war, December 8, 1941, Headquarters Company left our barracks at Clark Field forever. We loaded up with all of our combat gear and equipment and bivouacked in the vicinity of Clark Field under trees and camouflage netting. Obviously, our senior officers assumed that the Japanese airforce planes would return and bomb the barracks and supply houses, as indeed they did. We ate from field kitchens, having obtained rations and food from the Base Quartermaster at Fort Stotsenburg. We could move — men and equipment — in any direction on short notice. We were indeed combat ready.

From December 8 until December 25th, 1941, we repaired runways at Clark Field, which was frequently bombed. Our "Carryalls," bulldozers and tractors were busy all the time, not only at Clark Field, but later also at Dinalupihan, Cabcaben and Mariveles on Bataan, as the retreat to Bataan began to unfold. During the same period, we used our four- and six-ton trucks to move heavy artillery (155 mm G.P.F's) (World War I vintage "Long Toms") all over Luzon.

In the meantime, the Japanese had landed at Lamon Bay in the southern part of Luzon and in the north at Lingayen Gulf, at Aparri and Vigan, in the northern part of Luzon. General

Wainwright was responsible for repulsing the landings and attacks in the North. The highly vaunted (by MacArthur) Philippine Army Divisions "broke" before the enemy and disappeared in the bushes and jungles at the first sign of battle. They were untrained! Wainwright used mainly, as his reserve mobile force, the 26th Cavalry (Philippine Scouts) to engage the enemy. This famous horse cavalry regiment saved the northern force from annihilation by the Japanese by fighting a series of delaying actions from Lingayen Gulf south. This unit attacked the Japanese tanks with "Molotov Cocktails." The 26th Cavalry plus a platoon of light tanks of the 192nd Tank Battalion and some Philippine Scout artillery plus a few companies of Philippine Army troops that had held firm, fought a masterful delaying action. This planned five-stage retreat from Lingayen Gulf by Wainwright turned out to be the last horse cavalry action in the long and distinguished history of the U.S. Cavalry.

The 26th Cavalry plus the other troops mentioned above made possible the three plus months of the Bataan Campaign. Without their heroic efforts, the Southern Force, under General Parker, would never have made it to the Calumpit Bridge across the Pampanga River at San Fernando (Pampanga) and south into Bataan. These delaying tactics of Wainwright and his troops enabled us to bring several trucks to Stotsenburg, where we raided the full Quartermaster Warehouses of many cases of canned foods, between December 25th and 29th, 1941.

I recall making several trips, together with Sergeant Barry of Headquarters Company between December 25th and December 31st, 1941. We had to stop and take cover on several occasions due to bombing and strafing attacks by the Japanese.

Christmas Day, 1941

On Christmas Day, 1941, with the news being bad from both the northern front (from the landings of the Japanese at Lingayen Gulf) and from the Southern front (from the landings at Lamon Bay), we had to send officers and troops to Capas, Tarlac and Tarlac City to destroy the large gasoline dumps there. This could not be done by setting fire to them as this would have resulted in tremendous conflagrations, lighting up the entire countryside. The only thing that could be done was to knock two holes in each 55 gallon drum with a pickaxe. The officers and men were ankle deep in gasoline and suffered painful burns. They were oblivious to the fact that a single spark could cause a disastrous explosion. Insofar as this latter possibility is concerned, we were indeed fortunate.

On Christmas Day, Clark Field was evacuated. We rendered the runways at least temporarily inoperable by zigzagging across the runways with the trench dagger. We also destroyed the gasoline dump at Clark Field. We then began our retreat towards Bataan. War Plan Orange (WPO-3) was in effect. First we stopped at Dinalupihan where Company A was building an airstrip. Headquarters Company worked side by side with Company A to speed up the work.

About December 29, 1941, Headquarters Company moved deep into Bataan between Cabcaben and Mariveles at Kilometer Post, 165.5 (from Manila) at a place known as "Little Baguio." We were given the responsibility for improving and maintaining 30 kilometers of the East Side road including bridges, as well as the responsibility of building alternative bridges over many streams along the route. Headquarters Company also had the responsibility of supplies for the entire battalion — food, equipment and ammunition. We learned of the plans (under WPO-3) to blow up the Calumpit bridges over the Pampanga River.

Manila Revisited in the Last Week of December, 1941

Having learned of WPO-3 going into effect soon, with permission and wholehearted approval of the Battalion Executive Officer, I enlisted the indomitable Sergeant Barry and several volunteers and took several trucks into Manila to raid the abandoned Quartermaster Bodegas in the port area of Manila. Again we concentrated on tinned foods. On one of the return trips (we made several into Manila), we were strafed at Orani by Japanese fighter planes. We jumped into the ditches alongside the road and this time we suffered several casualties. It was slow going, not only because of the strafing attacks but also because of the fact that the road was filled with a large number of Filipino civilians retreating with American Forces into Bataan. It is really a wonder that we made it at all. Had the Japanese fighter planes returned one more time, they could have had a field day. The last foray that Sergeant Barry and I made into Manila was on December 31, 1941, just a few hours before the Japanese troop vanguard pursuing Gen. Parker's Corps arrived from the south. I have proof of my being in Manila on the last day of 1941 in the form of a receipt from the Philippine American Drug Co. (Botica Boie) at 95 Escolta, in the sum of 24.75 Pesos, which I carried in my wallet throughout the war.

We "beat it out" of Manila with our loaded trucks just ahead of the Japanese and just before General Wainwright ordered Colonel Skerry to blow up the big Calumpit bridges over the Pampanga River early in the morning of January 1, 1942.

Much of the food for which my men and I risked our lives was appropriated by the U.S. Army Quartermaster in Bataan for the central supply depot. Apparently we were not only enterprising, but foolhardy to return to Manila when most of our forces had abandoned it. This brings up the question of why the Quartermaster had not emptied the largest rice depot at Cabanatuan, as well as the many bodegas full of canned foods at the port area of Manila during the latter part of December. They certainly had the time and GHQ knew the plans.

Early 1942—Bataan

We rebuilt and maintained the Main Road (cobblestones) on the East Coast of Bataan from Km 194 to Mariveles at Km 174 using crushed gravel from the gravel pit at Mariveles, in addition to maintaining the road on the East Coast of Bataan, no easy task, inasmuch as it had been built by the Spanish about half a century earlier. We also moved and emplaced a number of large caliber naval guns — mostly eight inch bore — all along the coastline of the Bataan Peninsula from Bagac to Pilar.

On the West side of Bataan, there was a narrow unpaved road. We widened this and surfaced it with gravel. Unfortunately we had no binding material. But, it was an improved road from what we found there initially.

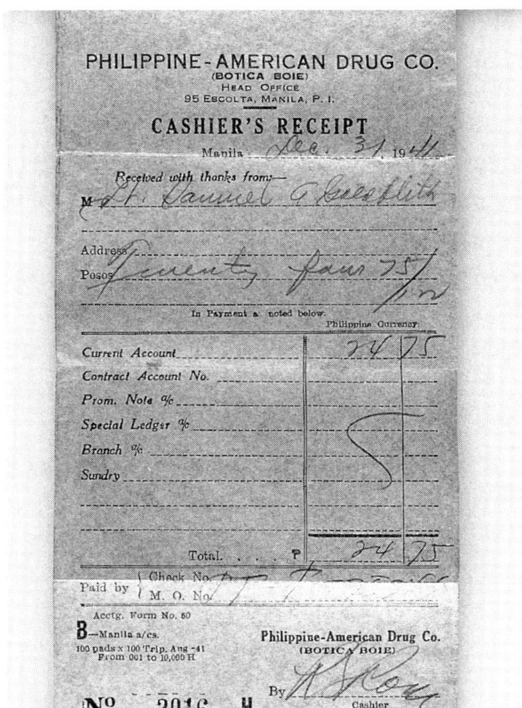

PHILIPPINE-AMERICAN DRUG CO.
(BOTICA BOIE)
HEAD OFFICE
95 ESCOLTA, MANILA, P. I.

CASHIER'S RECEIPT

Manila Dec. 31, 1941

Received with thanks from:—

M

Address

Pesos

In Payment a noted below.

Philippine Currency

Current Account	24	75
Contract Account No.		
Prom. Note a/c		
Special Ledger a/c		
Branch a/c		
Sundry		
Total . . . P	24	75

Paid by { Check No. / M. O. No.

Acctg. Form No. 60
B—Manila a/cs.
100 pads x 100 Trip. Aug -41
From 001 to 10,000 H

Philippine-American Drug Co.
(BOTICA BOIE)

By Cashier

Nº 2016 H

FIG. 3.1. RECEIPT FROM PHILIPPINE AMERICAN DRUG COMPANY SHOWING THAT AUTHOR WAS IN MANILA JUST A FEW HOURS BEFORE THE JAPANESE, DECEMBER 31, 1941

Author was in Manila raiding QMC Food Dumps.

Experiences in "Rescuing" An 155 mm Artillery Rifle and a Large Anti Aircraft Searchlight

The 803rd Eng. Bn., having a large number of heavy trucks, equipped with winches, was called upon by Bataan Force Headquarters, on numerous occasions, to pull large guns from place to place or to pull back large vehicles that had gone off the relatively narrow roads, and trails. On January 18th Major Fries, our battalion commander, was asked to send a crew with a heavy truck to recover a large artillery piece, a 155 mm rifle ("G.P.F.") that had been left behind by one of our artillery units in the retreat the night before from a line that ran across the Bataan Peninsular through Abucay to the Bagac-Pilar line. I volunteered to lead a group of two - three men with one of the larger trucks, equipped with a winch, to retrieve the artillery piece and return it to the unit, now in the rear of the Pilar area. We were given map coordinates of where the artillery piece had last been set up, near Abucay, and warned to be careful of Japanese as the whole situation was cloudy. Corporal Turturro and I and one or two others riding "shotgun" in the rear of the truck, proceeded to the map coordinate area, saw the gun and two of us began to hitch it up to the truck aided by the winch. Some Japanese spotted us and we had a bit of a skirmish, but recovered the gun (World War I vintage) and returned it to the artillery unit, now located behind the Main Line of Resistance, somewhere south of Orion. For this action, Corporal Turturro and I were cited for and later received the Silver Star for Gallantry in Action (General Order No. 5), 5th Interceptor Command, dated March 15, 1942, signed by Brig. General George, Commanding General. It had been an eerie and exciting night. While one artillery piece does not sound like much, it represented 1/16th of the heavy artillery guns on the east side sector of Bataan.[1]

Now being experienced in this sort of recovery, I was asked shortly thereafter to recover a large antiaircraft searchlight that had gone off the west side road just behind the Main Line of Resistance. This was about January 25th, 1942. I had a tractor, as well as a large truck with a winch. The problem was not in righting and moving the searchlight, but retaining access of the main road against the large numbers of Filipino troops retreating — literally Pampanga busloads — long enough to do so. My luck held, however. General Wainwright, the Commanding General of the Corps on the West Side of Bataan, had been inspecting the lines and just happened to come down in his command car with his staff. He asked me what the problem was. I told him and he ordered his staff to form a line with drawn sidearms and in an authoritarian voice, dared any and all to cross that line until the searchlight recovery operation had been completed. Thus he not only helped us accomplish our mission, but also stemmed the tide of retreat. The troops in the busses returned north to their units and I and my men returned the searchlight to its unit. It was the only time that I met General Wainwright in Bataan. I had met him at a cocktail party at Fort Stotsenburg before the war began. It was a memorable occasion, having a Lieut. General and his staff standing guard and holding the line for you.

It was at about that time that we received a message from General MacArthur that there could be no further retreat... "Thousands of men and hundreds of planes were on their way to relieve us" we were told. And, for a while, we believed it!

February - March, 1942

During February and early March, I was in charge of maintenance of the West Road from Km post 200-205. During this period, I was also in charge of demining the Binuagen Bridge and the bridge just above it. Together with Lieut. Everett Carney, we maintained the Mariveles cutoff road including gravelling and installing new treads on the bridges. This was an important road in that it went right by the main Q.M.C. fuel, food and ammunition supply dump of the Bataan forces.

Attempts to Assuage Hunger on Bataan

By February, 1942, we were down to quarter-rations in Bataan — of limited variety. These were mainly rice, canned fish and some meat. We were all losing weight, were short-tempered, hungry, etc. We knew that caribao had been slaughtered and delivered to us by the Quartermaster; The Quartermaster Corps had slaughtered 3,000-4,000 caribao and delivered such meat to the freezers of Corregidor.

I suggested to our mess sergeant that he detail a couple of men who knew how to butcher, I would get a small truck and we would go to the fields around Mariveles (at the foot of the Bataan Peninsula) and try to bag a caribao to augment our food supply. Off we went, one night in February, 1942, past Little Baguio, down the zig-zag road to Mariveles. We spotted only one caribao. I suggested to the men that we all take aim and fire together. I gave the command and we all fired our Model 1903 Springfields together and hit the poor animal. Then we struggled to get him onto the truck with a block and tackle and skids which we had brought along.

We returned immediately to the bivouac area where the mess sergeant and his men butchered the animal. For two days Headquarters Company of the 803rd Eng. Bn. had caribao stew and we all felt sated for a few days. I cannot recall which was tougher to chew, the caribao meat or the horse and mule meat from the 26th Cavalry.

In early March, Brig. Gen. Clinton A. Pierce, the intrepid commander of the 26th Cavalry (PS), informed General Wainwright that they were out of fodder for the animals and that he, Gen. Pierce, recommended to General Wainwright that the horses and pack mules of the 26th Cavalry be butchered and the meat distributed to the men on Bataan. Some 250 horses and 48 mules were slaughtered and the meat distributed to the troops on Bataan. We were in a static defense situation and there was really no need for horse cavalry in Bataan. The last horse cavalry action of the U.S. Army in war had been fought by these brave horses and men of the 26th Cavalry (PS) at the beaches of Lingayen and on the delaying actions during the retreat to Bataan. The horses and mules deserved a better fate!

General Wainwright agreed reluctantly, and with tears in his eyes, told his friend, General Pierce, to go ahead but that the first animal to be slaughtered must be "Joseph Conrad," General Wainwright's own personal horse — gorgeous animal and prize jumper. It must have hurt General Wainwright emotionally to have Joseph Conrad end his life in such a way.

During the latter part of March, Headquarters Company ran a school for newly commissioned engineer officers who had been civilian mining engineers in the northern provinces of Luzon and who had retreated with the American Forces into Bataan. However, the Japanese final offensives forced us to cut short the courses.

Fall of Bataan

There was increased aerial activity towards the end of March and our intelligence officers indicated that there was much more to come. We were put on call for Beach Defense. On Good Friday, April 3, 1942, the final Japanese offensive for Bataan began with intensive aerial bombardment with phosphorus/magnesium antipersonnel bombs. We suffered many casualties. It was a day none of us would ever forget.

On April 4th, we were ordered to discontinue all engineering work and to report to the Commanding General of the IInd Corps. About 400 of us, led by Captain Ingersoll, the Battalion Executive Officer, were given a sector of the rapidly diminishing front line which, by now, had moved rapidly south from the Mamala River to the Alangan River, about 3 miles north of Lamao. This was where the Main Line of Resistance was on April 8th, 1942, reinforcing the 57th Infantry, Philippine Scouts. We were now combat infantry and were fighting alongside the dismounted 26th Cavalry, (P.S.), plus some miscellaneous officers and men from a few other Philippine Scout units.[2]

In the middle of all this, proof positive of the fact that the administrative paper work was kept up at a good rate, I received my promotion to First Lieutenant, signed by Lt. Gen. Jonathan M. Wainwright and dated April 6, 1942 (Special Order #15, Hq. U.S.

Armed Forces in the Philippines). I was sworn in as a First Lieutenant by Jim Leggett, of the 803rd Engineers, on April 7, 1942. Needless to say that there was no partying. It was two days before the surrender of Bataan.

FIG. 3.2. BATTERY WAY AT CORREGIDOR - THE LAST BIG GUNS TO BE FIRED AT THE ENEMY
Commanded by Major William Massello, Jr. of Somerville, Mass.
Photo taken by author in December 1962.

We left the Main Line of Resistance, which had completely collapsed, by the evening of April 8th, 1942. By then we had retreated to Cabcaben with a M.L.R. of about 1500 men and were enveloped on both flanks. The grounded Air Corps troops had vanished from our left flank and the majority of Philippine Army troops on our right. We were told that General King had sent forward a white flag of surrender.

We returned to our bivouac area at Km Post 165.5, destroyed all of our heavy equipment and trucks, threw away the bolts of our Springfield rifles down a deep ravine, broke apart our .45's and scattered the parts all over and sat down to eat most of the C rations we had saved for a rainy day. Then we moved out to the main East Side road to await our fate. We had done all we could. We had fought to the end, were part of the last M.L.R. and now were about to become captives. I finally really understood the meaning of the Old Testament readings which told of the Hebrews going into captivity in Babylon.

END NOTES

[1] Extract of General Order No. 5 from Hq. of 5th Interceptor Command, Brig. Gen. George: ...Silver Star is awarded by the Commanding General, 5th Interceptor Command Luzon Force, United States Army Forces in the Far East for Gallantry In Action to: Samuel A. Goldblith (O-391719) in the vicinity of Abucay, Bataan, Philippine Islands, on January 18, 1942. Lieutenant Goldblith, with one enlisted man, went to a place near Abucay, and brought back a 155 MM gun. Disregarding the fact that the enemy situation was obscure in that vicinity, Lieutenant Goldblith unhesitatingly went to Abucay under cover of darkness, secured the gun, and delivered it to the Philippine Ordnance Depot, thereby accomplishing the desired mission. Next of Kin: Diana L. Goldblith, wife, 24 Holden Street, Lowell, Massachusetts.

The Bronze Star Medal was awarded (after the end of the war) to Captain Samuel A. Goldblith, United States Army for "meritorious achievement in ground combat against the armed enemy during World War II in the Asiatic-Pacific Theatre of Operations." The 1st Oak Leaf Cluster (to The Bronze Star Medal) was awarded for "meritorious achievement during combat in the South West Pacific Theatre of Operations from December 7, 1941 to 10 May, 1942. The actions of Captain Goldblith were in keeping with the highest traditions of the military service and reflect distinct credit upon him, his unit and the United States Army." s/by the Secretary of the Army and the Adjutant General, 4 May, 1994.

[2] For this activity, we were to be entitled to wear the Combat Infantry Badge.

CHAPTER 4

THE BATAAN DEATH MARCH AND EARLY DAYS OF CAPTIVITY (APRIL 9 - APRIL 22, 1942)

The Bataan Death March has been described in many books and articles — some by participants and others who are historians. Thus I will not attempt any detailed description other than personal reminiscences that were significant to me. Among the books are *Surrender and Survival* by E. Bartlett Kerr, 1986, published by Wm Morrow & Co., New York, N.Y., and an historical fiction account by John Toland entitled *Gods of War*, published by Doubleday and Co., 1985, and paperback published by Tom Doherty Associates, Inc., N.Y., 1986, 598 pp.

On the evening of April 8th, 1942, at about 9:30 PM, the main ammunition dump on Bataan, located near one of the two main field hospitals, was blown up. It was like a huge fireworks display of tremendous force, noise, lights and color. Debris fell everywhere for several kilometers around the dump. I should also mention that at this time, we at Headquarters Company, in addition to destroying all our heavy equipment and trucks as related earlier, destroyed all the battalion records, including all unaccountable overages in foodstuffs and equipment dating from our arrival in Stotsenburg until the beginning of the war. The records were thus "sanitized."

At the Main Road, we had been instructed to surrender to the nearest Japanese officer who showed up. After a period of several hours in the hot sun, we were herded by the Japanese soldiers who came along, into columns and were slowly (at first) marched northward on what has since been known as "The Bataan Death March."

Japanese troops in trucks and on foot passed by us, heading south to position their artillery on high ground to bombard

Corregidor. We were jabbed at with rifle butts by troops in the trucks as they drove by. Periodically we had rings, watches and eyeglasses taken off us by the foot soldiers marching into Bataan.

My M.I.T. class ring was ripped off my finger as was my wrist watch. My eyeglasses were broken by a whack on the head. The dust and heat were oppressive and stifling. Large 240 mm Japanese artillery pieces passed us. They were the largest mobile artillery guns that I had ever seen.

When we began the march north at Km. 165.5, most of us began with an extra uniform, mess kit, canteen, towels, extra underwear, raincoat, etc. As the heat of the day became more oppressive, our thirst unquenched, our strength began to ebb. I (and many others) shed all that I was carrying except the clothes on my back, my canteen and mess kit and a pink towel (from Diana's trousseau), which I had brought with me from home. I wore it over my head as a shade from the sun and, whenever possible, dipped the towel into water in ditches along the side of the road to help me cool my head.

I recall passing through and/or stopping at Orion, Lamao, Balanga, Abucay (where in January I had retrieved the 155 mm G.P.F.), and Hermosa. I marched with our Battalion C.O., Capt. Herbert Ingersoll. At some stops, we stood in line at a single spigot for what seemed like an eternity, to obtain a canteen full of drinking water. We first stopped at Cabcaben Field (that we had built) the first night. No food. On the night of the 11th of April we arrived at Limay. Captain Ingersoll and I had a chance to head north by truck, but we sent those in worse condition than we were. We continued our trek northward reaching Balanga about 5 PM on April 11th. Much shooting was going on. We were lucky. We reached Orani about one AM on the 12th. We were herded into a field. There was room only to sit; not to lie down. On the 12th of April we received two small meals of rice, (about a mess kit full). On the 13th, we walked through Hermosa and reached Lubao about 11 AM. At Gargua, Captain Ingersoll and I somehow got a carametta (I believe he paid a Filipino) and drove on the buggy to about Km83. Enroute, we were strafed by a Japanese airplane — another close call. We finally reached San Fernando, Pampanga Province about 4 PM on the 13th.

At one of the places mentioned above, I've forgotten which, we received a handful of rice. There were open cesspools and latrines everywhere with disease-bearing flies and maggots all around. Those who lingered or left the line of march were shot or bayoneted. At each watering stop, I recall taking one of the small tubes of tincture of iodine from my WWI first aid kit and dosing the water in my canteen. The iodinated water tasted foul but was probably safe. I had remembered from my 5.01 class in chemistry, that iodine, chlorine and bromine were in the same section of the Periodic Table of the Elements. Moreover, the knowledge gained in Municipal Sanitation from Prof. Horwood probably saved my life. I avoided one of the scourges of the Death March — dysentery.

The fatigue was terrible. I recall that at one rest stop in the middle of the night, we sat down on the road to rest and almost did not get up. I had reached the point of not caring any longer. But, Captain Ingersoll, with whom I was still trudging along, just about dragged me up and forced me to continue. Thanks to Herb and the Good Lord above, I made it. I suspect that The Death March was about the worst experience of my life — even worse than the trip to Japan from Manila, the following November, 1942, aboard the Nagato Maru.

At San Fernando (Pampanga), after a bit more sun treatment, on April 14th, we were shoved aboard boxcars, one hundred men per small car. The doors were closed and locked. There was no ventilation. In these boxcars we were taken from San Fernando to Capas, Tarlac, where those of us who had survived the stand-up, suffocating rail journey, were allowed to disembark. We were lined up and marched to Camp O'Donnell. Along the 7-mile route, thoughtful Filipinos had placed pieces of sugar cane and bamboo containers of water which we were permitted to take — on the run.

Camp O'Donnell — First POW Camp

Camp O'Donnell was an abandoned and unfinished Philippine Army training camp. It had a series of bamboo and nipa constructed, single story barracks with open, unscreened ventilation, no mosquito nettings, sanitary trenches for facilities; inadequate water; no washing facilities; and inadequate food.

Our introduction to Camp O'Donnell was when, immediately on arrival after the seven mile march from the Tarlac train depot, we were lined up and subjected to a most horrible speech by the camp commandant, a Captain Yoshio Tauneyoshi, a reservist in the Imperial Japanese Army. He delivered a terrible tirade against America and told of all the Japanese victories against the U.S.A., Great Britain and the Netherlands — the fall of Hong Kong, Indonesia, Singapore, etc. We were told, in no uncertain terms, that we were not Prisoners of War but Captives! We did not surrender but rather we were captured. Japan, we were told, would fight to win — 100 years if necessary. All this long tirade took place in the broiling sun, while we were unfed and dehydrated.

Then we were stripped and searched again and finally we wandered into the camp. On one side of the road was the enclosure in which the Americans were kept and on the other side of the road, the Filipino encampment.

The food was a watery rice gruel called "lugao." It tasted terrible. However, it was that or nothing, so I felt that if I wanted a chance of survival, I had better eat it. The water supplies were scant. A single faucet where we would go to fill our canteens. Thus, life went on for several days — lugao, boredom, disease — malaria, malnutrition of all sorts and men dying in large numbers, daily. Over 1,800 Americans and 28,000 Filipinos are buried at O'Donnell. It was one of the worst of the Japanese horror camps.

In the midst of all this, — less than a month after the surrender of Bataan, the Japanese Camp Office asked for engineering officers and men to go on a work detail in the northern provinces of Luzon. I volunteered. I reasoned that it could be no worse than where I was, since we were losing more than 100 men per day to malaria and dysentery, mainly. The date was April 22nd, 1942. We were to leave the next day.

CHAPTER 5

THE MOUNTAIN PROVINCE DETAIL (APRIL 23, 1942 - CIRCA AUGUST 1, 1942)

Camp O'Donnell to Mountain Province

With trust in the Almighty and hopes that we had made a wise decision to leave Camp O'Donnell, we were transported by trucks on April 23rd, 1942 — heading north. We were split up into three groups. The first under Captain Robert Chandler — mainly men of Company C of the 803rd Engineers, with Lieut. Ralph Gibbs, also, went to Baguio (the "summer capital" of the Philippines) and worked on roads and bridges in the Mountain Province. The other two groups — one under Captain Eugene Radcliff, and the other under Captain William "Sorry" Thomas and Captain Peter Reynolds and myself numbered 200 men. We headed for San Fernando (La Union Province).

We were then split into two groups. The larger with Captains Reynolds, Thomas and the writer numbered 150 men. The second group was 50 men under Captain Radcliff.

The story herein relates to the group under Captains Reynolds, Thomas and myself. We were attached to a Japanese Engineer unit. Our men were split into groups of three, four, five and ten men who acted as "dog-robbers" for the various Japanese squads. We proceeded north on Route 3 (the same route that went through Angeles, Stotsenburg-Clark Field) to about 43 kilometers west of Cervantes, where a big bridge had been blown out. From there it was on foot with the Americans acting as "cargidores," for the Japanese and carrying large loads of material for the Japanese engineers. Some of us were carrying up to 135 pounds all up and down hill in the mountains with a minimum load of 50-75 pounds each.

We rested for three days at a barrio (probably Butao) about 30 km west of Cervantes. I'm sure that the reason for the stay was not consideration for the POW's, but rather because of the fact that the Japanese needed to make plans for their next moves. Later we learned that they were strategizing on how to capture Lt. Colonel John P. Horan and his group of "irregulars." Col. Horan had been C.O. of Camp John Hay of Baguio. Colonel Horan, according to the official history of the Defense of the Philippines, had kept MacArthur informed as to Japanese movements in the Mountain Province while at Camp John Hay. On December 23, 1941, his last radiogram to MacArthur, with the Japanese advancing from all sides, read: "My right hand is in a vise, my nose in an inverted funnel, constipated my bowels, open my south paw." (from L. Morton, *The Fall of the Philippines, U. S. Army Military History,* 1985, pages 135-6).

Apparently, the Japanese engineering detachment that we were the cargidores for, was attempting to reopen the mountain roads and capture Col. Horan and his men. Horan's force was augmented by a number of mining engineers from the area. Using guerrilla tactics, they harassed the Japanese. It is to be noted that Col. Horan and his irregulars conducted 51 hit-and-run skirmishes against the Japanese regiment for which we were the cargidores. The Japanese were well-equipped with machine guns and had one 47 mm cannon, which we had to carry, broken down, of course. At Cervantes, for example, we were delayed for two days by heavy rains which had swollen the river. Lt. Mel Harrison (I learned later at Cabanatuan POW Camp), had his outpost setup across the river and the Japanese sprayed the area with machine gun fire. I know that we were lucky not to have been hit in the crossfire. Col. Horan died on July 31, 1985 at age of 93. He and his men never surrendered until ordered to do so by General Wainwright when Wainwright surrendered Corregidor and all of the Philippines.

In my own group of "dog robbers" were Master Sergeant Dickenson, U.S. Engineers depot, Staff Sgt. Shlingerhoff of Hq. Co., another staff sergeant of USSAFE Hq., whose name I do not recall, and Sergeant Hogan of Company B.

At Cervantes

Somewhere along the route, we "procured" a caribao cart (minus the caribao). It was rugged mountain terrain and the going was tough. There was no distinction made between officers and enlisted men, as we all pitched in together. It took us three days to make 39 km because of the heavy loads we were carrying/dragging. In addition we had to cook and wash up for all the squad. At Cervantes we remained for two days due to rains. We crossed the river by cable at night (a most interesting experience). In Cervantes, the village priest passed out cigars to all. I recall getting quite nauseous. The dark cigars were not made for my stomach. The Japanese had poor sanitary practices. They also raided many of the abandoned houses, taking whatever suited them.

After we crossed the river, we proceeded east to Bauko, on to Rte. 4 marching uphill for about 6 km, when we rested at 3:00 AM. We stopped at a small barrio and then went on further uphill for 6 more km and finally downhill for about 8 additional km when we arrived at Bauko (enroute to Bontoc). We arrived at Bauko in the afternoon of the next day — a long and hard trek. We remained at the Bauko schoolhouse all night. There Japanese looted all the supplies, gardens, garden stores, etc., forcing us to aid them. They shot pigs for supper — nothing was spared in the deserted barrio. Early the next morning we continued east to Sabangen — a distance of about 12 km. Here, many of the people had not left, but had taken their most valued possessions to the surrounding mountains and returned to the village. The Japanese still shot what they wanted for food — mainly pigs and looted hen houses for eggs. At Sabangen, the village priest and the nuns (Sisters of St. Augustine, a Belgian Order) gave medical aid to us (which the Japanese permitted), as well as brown sugar which we were very thankful for. The drugs they gave us, (mainly quinine), as well as first aid attention, were not only very much appreciated by us but also probably saved many lives.

To Bontoc

The next day, we left on foot for Bontoc, a distance of 19 km. This was quite a rugged hike due to the fact that the Japanese got some additional supplies from Baguio while at Sabangen and also purchased some rice there. The next morning, after we had arrived at Bontoc, some Filipinos came to the Japanese officer in charge and informed him that an American was found dead by the side of the road, about half a kilometer back. I went back to the spot with some of the men and found Sergeant Hogan of Co. B, 803rd Engrs., one of our group, who, ill all the way with malaria, got separated from us and although he was without a load, fell behind. The "report" we had received, via the grapevine, was that he had been hit on the jaw with a rifle butt by a Japanese and had died there.

When I went to the spot, there was foam on his mouth and a Bible in his hands with his rosary beads. Rev. Father Miguel Veys, the Bontoc priest, conducted services and we buried him in the Bontoc Cemetery.

The Japanese began to loot Bontoc. But the Igorots, small in stature but very brave and strong, decided to give the Japanese some gifts and the looting ceased — especially with the arrival of a higher ranking commanding officer who forbade such practices. We remained near Bontoc for about two weeks, repairing a bridge which Colonel Horan and his men had blown. I helped to design the bridge but did little work on the bridge due to some sort of infectious diarrhea that I had picked up. Then, I and three or four other American "dog-robbers" from other squads went to Bontoc where we remained and we were treated reasonably well.

Bontoc to Lubuagen

At this point, the Japanese regiment to which we had been attached had been ordered to depart for Vigan and thence to Japan. All of the Americans, about 125 of us, under Captain Reynolds, Captain Thomas and me, began work repairing the bridges on the Bontoc-Lubuagen Road, a distance of about 68 km. We carried timbers to the bridge sites and kept moving up the road working on repairing the road surface as well until we reached Mamangot, where we stayed for two weeks. I should pause in this narrative here and point out that the road "hangs" on the side of mountains, is often washed out by rock slides and is in need of constant care. It is gorgeous scenery with the famous rice terraces all around. Below there is a deep gorge with the swiftly flowing crystal Chico river at the bottom of the gorge. From Mamangot, we carried supplies up and down two very steep gorges where the bridges had been blown up. Here, many of my fellow American POW's developed malaria and dysentery and were sent back to Bontoc, the hospital at Baguio and then to the concentration camp. At Tucucan, about 6 km outside of Bontoc, we were asked if we wanted to go back to the concentration camp at O'Donnell and we replied in the affirmative. We were told that we could go back in two weeks. The two weeks would have been up by June 1st, 1942. But, in the event, the two weeks stretched until August 1, 1942.

Each day we carried supplies for about 10 km in each direction, walking 20 km each day. We were issued rice and salt only and we pilfered sugar, flour and dried fish from the Japanese. At Mamangot, Captain Thomas contracted malaria and dysentery and was sent back to Bontoc. At 12 km north of Bontoc, we met Captain Radcliff and his men who had come in the opposite direction from Lubuagen, also acting as cargidores. They were attacked by Filipino-American forces and several Americans from Company B had been killed and several others were wounded. Here, Captain Radcliff developed malaria and went back to Bontoc and Baguio.

From Mamangot we marched to Basao, a distance of about 10 km, stayed overnight and then rode in Japanese trucks to Lubuagen on the following day, a distance of about 35 km.

We stayed at Lubuagen for four days — my diary shows this to be about July 1, 1942. Seventy-three of us were left somewhat "healthy" out of two hundred. The remainder had been sent back, died or were killed. From Lubuagen, two Americans were sent back to Bontoc Hospital.

Some of our group (about 40 men) under Captain Reynolds were sent to work at the Baton-Bouhi mine, 20 km outside of Lubuagen. At Lubuagen we were housed in a small schoolhouse known as Pud-pud. Here 13 of us, including me, were ill — some seriously. The Sisters of St. Augustine were very good to us all over the Mountain Province, but especially here in Lubuagen and Bontoc. They were brave in bringing us quinine, other medicines, anti-dysentery medicines, bandages, etc. They brought us food and drugs and sent food with some of the native boys. In my mind I can still see the nuns walking, half-running along the tops of the rice paddies to bring us medical supplies and cooked food, cleaning our festering sores. They were very brave ladies, indeed, as well as kindhearted.

Sister M. Arthur (Sr. Jean Delbeche)

It was here at Lubuagen that I met Sister Marie Arthur, that wonderful nun from Belgium, who was so brave and did so much to ease our plight. I'll write in more detail about her later on in this memoir. Meanwhile, my shoes had worn through and were held together by cloth, rope, etc. My feet certainly took a beating. On August 1, 1942, we assembled together with Captain Reynolds and his men and marched back as a group to Bontoc — I, by now, shoeless. But I was not the only one. It happened to all of us. We had climbed up and down mountain trails, rocky roads, etc. It took us three days to march from Lubuagen to Bontoc. (In 1975 I learned from Sr. Jean Delbeche (Sr. M. Arthur) that earlier that year she went by bus to Lubuagen from Manila to celebrate the founding of the school in Lubuagen by the Order. The journey took a total of three days. One day to Baguio, one day from Baguio to Bontoc and one day from Bontoc to Lubuagen.

At Bontoc, some of us (with small feet), were issued shoes and most of us received denim clothing. However, people like me with relatively large feet (size 10), still had to go on shoeless. We remained overnight in Bontoc and thence went by truck to Baguio where we remained two days. Baguio is a gorgeous city with beautiful scenery, as is all the Mountain Province. In Baguio, we met Colonel Horan's officers, as he had surrendered, by then, under orders from General Wainwright. We all proceeded by truck and train to Cabanatuan Prison Camp.

We often hear the remark to the effect of never volunteering in the Army. Was I correct in volunteering for the work detail? Would I have been wiser to stay at O'Donnell? We'll never know. In retrospect, I was probably wise in volunteering, inasmuch as it was a more temperate climate in the Mountain Province, and the work was very hard, but probably all right for me — undernourished yes, but other than malaria, in reasonable health, lean, and for me, hard. In O'Donnell, there was nothing to do but lie around, amongst the flies — physical and mental decay. It is really a marvel as to what the human body can take in terms of punishment. We had been up and down mountain roads, carrying loads often as much as 135 to 145 pounds — rifles, machine guns, bags of rice, Japanese packs, 47 mm guns, and carrying heavy timbers, etc.

Because we were interspersed among the Japanese, Colonel Horan's men, having been told of our presence, by the natives, did not dare to engage the Japanese in shooting for fear of harming us POW's. We were lucky to reach Cabanatuan, more or less in one piece, but alive. Later during the war, in another POW camp, we had learned that the death rate at O'Donnell while we were gone had been over 100 men per day!

The Catholic Priests

After the war, I wrote a letter to Rev. Fr. Miguel Veys of Bontoc, dated November 12, 1945 to thank him for all he did for us. I learned in a letter from Rev. Fr. Leon Quintelier who succeeded Fr. Veys in August, 1942, by his letter to me of Nov. 29th, 1945 that six American POW's had been buried in Bontoc Cemetery or behind the hospital in Bontoc and that each November

1st, the natives of Bontoc clear the gravesites of their loved ones and the priest blesses the graves. On Nov. 1, 1942, Fr. Quintelier blessed the graves of the Americans buried there after the gravesites had been cleared and flowers placed thereon by members of his church. He informed me that because the graves of five Japanese were not cleaned up and blessed by Fr. Quintelier, he was put into solitary confinement in a dark, "bad-smelling room" and later put into one month's solitary detention in his room at the convent. Poor Fr. Quintelier, he too suffered and because of us.

I had given Fr. Veys, to keep for me, my original citation for the Silver Star and the Certificate attesting to my promotion to a first lieutenancy. Fr. Veys had given the documents to Fr. Quintelier who had turned them over to some American Lieutenant, only one week before he received my letter. I never saw the originals again. No matter. The War Department had records of both matters and in due time, I received copies of both documents.

At Cabanatuan, we were welcomed by 803rd Engineers who had survived the horrors of O'Donnell, I received a pair of shoes that fit and settled into the daily routine at Cabanatuan. Thanks to the Good Lord, the Sisters of St. Augustine and luck, we survived the work detail in Mountain Province.

CHAPTER 6

CABANATUAN POW CAMP (AUGUST - NOVEMBER, 1942)

Cabanatuan POW Camp

When we came down, by truck, from the Mountain Province work detail, we were taken to Cabanatuan Camp No. 1. Cabanatuan City is located in Nueva Ecija Province, about 100 miles north of Manila. It had been the site of a United States Agricultural Experiment Station. When the 91st Philippine Army Division had been mobilized, it was housed in Cabanatuan in three separate camps, each designated for one of the three regiments of the division. Each barracks was constructed of bamboo, nipa-thatched, swali-sided and was 60 feet by 40 feet and built to accommodate forty Filipino soldiers in each barracks. As POW quarters, each building accommodated 120 POW's.

Cabanatuan POW Camp No. 1, to which we were taken, was about 600 by 800 yards in area. There was eight-foot-high barbed wire around the camp with a number of four-story guard towers at even intervals along the fence. Camp No. 1 housed most of those who survived Camp O'Donnell. The following statistical estimates have been made to give the reader some idea of the magnitude and numbers involved and the toll that the Death March and deaths due to diseases, such as malaria, dysentery and malnutrition (See Knox, Donald — *Death March* — Harcourt, Brace, & Jovanovich Publishers, 1981, New York, 482 pp).

Camp No. 3 housed the sick and wounded who had remained behind on Bataan in the two field hospitals as well as, a few weeks later, the men who had surrendered on Corregidor, until the total in Camp No. 3 reached 6,066 POW's. The remaining Corregidor prisoners, housed in the interim at Bilibid Prison in Manila, some 1,500 men, were placed in Camp No. 2, arriving about May 31, 1942. Since there was no water, they were moved to Camp No. 1. Beginning June 2, 1942, because of the heavy death toll among the prisoners and because of water shortage in O'Donnell, the men from O'Donnell were transferred to Camp No. 1.

According to Knox, about 650 men died on the Death March and about 1,500 men died at Camp O'Donnell. About 2,000 were assigned to work details, such as our group in the Mountain Province. It is estimated that about 5,850 men were transferred directly from O'Donnell. These were surely the sickest, most weak and disabled of the POW population due to malaria, and the effects of the Death March, etc.

In June, 740 men died at Cabanatuan, with 130 alone from an outbreak of diphtheria. We were spared exposure to this dread disease by being in the Mountain Province, even with its risks and hazards. By 1942, according to Knox, nearly 2,000 men died in Cabanatuan Camp No. 1, where we were! By comparison, Knox points out, only 61 died in Camp No. 3, where the men were relatively more healthy. This gives one a picture of the state of morbidity and mortality when we arrived in Cabanatuan, Camp No. 1. I recall that one day after our arrival, I was on the burial detail, and we buried over one hundred men!

The crude hospital wards in Camp No. 1 housed about one-third of the camp! The American POW physicians attempted to limit infection from dysentery patients by isolating them from the other patients. They were in what was called "Zero Ward" — expected to die — chance for survival approached zero. At O'Donnell, this ward was named "St. Peter's Ward." The only type of medicine available was really only charcoal from the kitchen. The physicians had little to work with either at O'Donnell or Cabanatuan.

Camp 1, Cabanatuan

On arrival at Cabanatuan, our detail was assigned to Group III in Camp 1. The kitchen for Group III was a separate unit known as "Fitch's Kitchen" with a sign over the front which read "Fitch's Kitchen Quit Your Bitchin'." Major Fitch had been a QMC officer and being in charge of a kitchen was a plum job in a POW camp. Some prisoners had money — dollars and pesos — and could purchase local foodstuffs from Filipinos, who brought camp supplies via caribou carts, and who took messages out and returned with replies. Other sources of communication and thus goods and money were the work details and truck drivers. The underground communications link and "thieves market" was a tribute to the resourcefulness of the prisoners. Moreover, the Filipinos often took their own lives into their hands by helping the Americans.

I was fortunate in having a classmate in Manila, Oscar B. Mapua, who sent me some pesos for which I shall be eternally grateful. I used them to purchase quinine tablets which I used, as well as distributed to some of my fellow 803rd officer POW's. I may have purchased some food but I doubt it, on reflecting back. One was hungry all the time, anyway. While we were not getting enough to eat in terms of calories, vitamins or protein, we desperately needed quinine. I also sprouted mung beans to get vitamin C and chewed on carefully washed grass, which I reasoned if it had chlorophyll, it must also contain vitamin C. The two meals per day consisted of rice in a pasty form known as "lugao" for breakfast and steamed rice supplemented with some type of native greens for lunch and supper. Everyday survival was a struggle, as was the danger from inactivity and boredom. I knew that we could not survive long under the conditions in which we were existing.

We had prisoners in the camp and in our barracks from all over the U.S. We had songfests periodically with entertainment by POW's. It was here that I learned the college cheers of Kansas, Rock Chock KU; or Virginia Polytechnic Institute, Ray Ri VPI; "The eyes of Texas are upon you," etc. They bring back memories of starry lit skies, and hunger assuaged by pleasant evenings, every now and then. We had some amazingly good amateur entertainers. Lectures were not popular in Cabanatuan — we were just too tired and hungry to absorb food for the mind; only food for the body. I even went to a number of Sunday Protestant worship services and still remember the hymn "The Church in the Wildwood." The quality of the entertainment was not important, we would listen to anything light and entertaining.

Prime Minister Hideki Tojo's No Work - No Food Policy

In October, 1942, we learned of possible movements of the prisoners to other places both in Japan and elsewhere. Later we learned that this was as a result of Prime Minister Hideki Tojo's decision to transport some POW's on each vessel returning to Japan from Manila to help with the acute labor shortage in Japan, Manchuoko, and south in Mindanao. His proposal was a no-work, no-eat policy based on the national problem of shortage of food and wartime emergencies. Tojo's policy was announced on May 30, 1942 at Zentsuji where he gave a talk to the commander and staff at Zentsuji. He stated in part: "The present situation of affairs in this country does not permit anyone to lie idle doing nothing but eating freely..." (from Kerr, E.B., *Surrender and Survival*; William Morrow and Co., N.Y., 356 pp).

Cabanatuan to Manila

On November 5, 1942, at about 03:00 hours, we indeed left Cabanatuan Camp. No. 1, and walked to Cabanatuan City to the freight yards. We had another speech, this time by an elderly, wizened Japanese Colonel who told us that "we must guard our health." How many times was I to hear that statement while a POW? He informed us that we were indeed going to Japan. There we boarded the dreaded box cars bound for Manila — 98 per car plus one or two armed guards in each car. This time, unlike the previous trip from San Fernando, Pampanga Province to Capas, Tarlac, the doors of each freight car were left open! Even so, doors open and all, with the tropical sun, one gets a good idea of what a trip in hell must be like!

Regardless of whether or not we stopped at a station, all along the route, Filipinos had set up the "bamboo network" and were on hand to throw food to us into the freight cars, Bless them. They had little enough themselves. At one station, they even brought buckets of fruit juice drink, as well as a fruit-rice pie.

We reached Manila late in the day, around 17:00 hours and were marched, with our meager possessions and tattered uniforms (of sorts) right through the business area of Manila. We were sure that it was done to humiliate us. Our former comrades-in-arms, however, were all along the route and held their hands over their faces with many making a "V" sign with their index finger and the middle finger, which they had learned during the brief but intense battles. It was a heartening sign and meant much to us.

We spent that night, the night of November 6-7, 1942, on a cement floor near Pier 7 — the same pier where we had arrived to the accompaniment of a brass band, the previous October — a shade over one year earlier. So much had transpired since then! There was quite a supply of uncured tobacco leaves in that old warehouse and many of the officers and men, the inveterate smokers amongst us, helped themselves to some. We watched the loading of an old freighter by Filipino and Japanese workers all the next day, and at 17:00 hours on the 6th of November, we were herded like cattle aboard this old freighter — probably in use since the turn of the century, "The Nagato Maru."

The Nagato Maru — Manila, P.I. to Moji, Japan

We were divided into three holds on small decks that had been constructed between each regular deck. The top deck housed the regular Japanese troops who were being returned to Japan with the uppermost deck being reserved for the Japanese officers together with "futons" to lie down on and fresh fruit for them to munch on.

We were crammed in like pickles in a jar. Not enough room to lie down, perish the thought. Barely enough room to sit, with knees drawn up. There were 1,500 POW's and probably a like number of Japanese — with more room of course. Every time I see an old movie with slaves at the oars, I am reminded of our trip aboard the Negato Maru.

If anyone wanted to turn over or change position, it was a project to have the people around you to do similarly. It was hot in the tropics — insufferably so in the holds of our boat. The toilets were at the aft end, hanging over the port and starboard sides of the vessel. Thus no running water was needed for toilets! We simply fed the ocean directly! But to use these, one had to climb up, get permission and race to the rear of the ship. Many of the POW's had dysentery or at least a good case of the "trots" and could not make it. So we had "honey buckets" in each hold for those who could not climb the ladders. The body lice on those that had them spread very quickly so that soon all of us were scratching the "cooties" and would do so for about another year until we reached Zentsuji.

The honey buckets would go up to be emptied over the rear sides of the vessel and food would come down twice a day — rice and a watery soup. We cleaned our mess kits by eating up all of the food (bad as it was) and then wiping the mess kit. The result, obviously, was an increase in the number of cases of diarrhea.

Men with the trots had to sit near the buckets. When dysentery hits, one has to really go and there is no time to move any distance. Moreover, the more one goes, the weaker one becomes. That was life in November, 1942. Added to this was the fact that often we were kicked or beaten by the Japanese soldiers enroute to the latrines. Then, there was always a line at the "heads" and one had to wait one's turn.

On November 11th, the Nagato Maru reached Takao, Formosa. Only a few men were permitted to go onto the top deck. We remained at Takao until November 15th when we left for Mako, Formosa, where we remained until November 18th. Some were told that a bad storm had delayed us.

Somewhere between Mako and Moji, Japan, there was a great deal of excitement. The ship, of course, zig-zagged its way north but suddenly, it changed its course and the cannon (probably a 75 mm, as I recall it,) at the bow of the vessel, was fired, having been aimed by changing the course of the ship. One of the escorting destroyers fired some depth charges at what we supposed was a submarine. We never found out what actually happened. Suffice it to say that if a torpedo had hit the Nagato Maru we, in the holds, were goners.

By then, it was late November. The weather was getting colder as we headed northwards. We began to shiver and were very cold. But the air down in the holds smelled better. The Nagato Maru and assorted passenger cargo pulled into Moji, Japan on November 24th, 1942 at about 17:00 hours. How the human body could take such abuse sitting in one spot, hardly ever moving, if one was lucky enough not to have the "trots;" insufficient food, lice, etc., etc. Now the intense cold as we disembarked. We had lost ten men during the trip. It's a miracle that many more did not die.

"Get the Lead Out of Your Asses!"
Shimonoseki to Osaka by Train

Suddenly, as we were being herded along the pier where we had disembarked, a voice came out of the darkness shouting, in perfect Americanese, "Come on now, move!; get the lead out of your asses!" We learned later that the name of the interpreter was Fujimoto who was dressed like he had just come off Madison or Fifth Avenue. We were marched to a ferry boat which took us from Moji to Shimonoseki, where we were given a "bento," a small wooden box, containing rice, pickled daikons, seaweed, and a slice of tangerine. We ate in our third-class seats on a train to transport us to Osaka, where we arrived on November 26th, 1942; Thanksgiving Day in the U.S.A.! And for us also, it was a Thanksgiving for the Good Lord had seen us, who had lived, safely to Honshu, the main island of Japan.

Later on that day, after sitting around Umeda Station for a while, our group of about 450 officers and men were marched to a

POW Camp known as "Osaka Umeda Bunsho." Of the three camps in the Osaka area where the American POW's from the Nagato Maru were taken, namely: Yodogawa, Tanagawa and Umeda, the Umeda camp is said to have been the worst of the lot. At Umeda, we lost almost 25% of our group the first winter in Japan, 115 officers and men out of a total of 458, because of dysentery, beri-beri, malaria and pneumonia. The basic underlying reason was severe malnutrition.

Our group was led by Lt. Col. G.L. Fields, who died on January 24, 1943. Lt. Col. Theodore Wickord of the 194th Tank Battalion became the POW Commander at Umeda. Col. Fields had a brother in Lowell, Mass, and I saw his brother and sister-in-law in Lowell after the war and informed them of the circumstances of Col. Field's death.

CHAPTER 7

OSAKA UMEDA BUNSHO (NOVEMBER 1942 - AUGUST 1943)

The Camp

Rations at Osaka Umeda Bunsho POW Camp were poor. The officers received less than the enlisted men and the latter did not receive enough food to survive on. The camp commandant, a ranking sergeant, was always after the officers and trying to drive a wedge between the officers and enlisted men. We were in a three-story wooden building with a tile roof. It had a kitchen and a Japanese Guard Quarters. A large bath was located in a small adjacent building for communal bathing once per week on a Saturday night. It was a chance to get warm by a cold water wash outside the tub, Japanese style, and then a ten minute hot water soak.

The upper two floors were divided into large rooms with raised platforms on which were tatami mats for sleeping and resting or sitting. The additional complement to our sleeping, other than thin blankets, was a hard, small, rice- and straw-filled pillow, such as one sees in the old Japanese wood block prints of women and their men friends. Our meager possessions were on a shelf above our heads. The heat was practically nonexistent. There was a charcoal brazier in each room which was used two hours each evening when it was extremely cold. The charcoal was very scarce. The enlisted men were separated from the officers, and the officers of field grade rank were in a room away from the more junior officers. I was in a room with first and second lieutenants. The whole camp was small and in a relatively small area of about 150 feet by 200 feet. One day one of the prisoners who understood some Japanese overheard the sergeant say to the corporal of the guard shift, “Go upstairs and see what the officers are doing and stop them.”

Rations were insufficient both in quality and in quantity when we arrived at Umeda and became worse in March, 1943, when the food supply for the camp was taken over by the Nippon Tsuun Kaisha Co., a freight handling and forwarding company much like the old American Railway Express. The work was hard and the days were long. We handled all sorts of freight from cars loaded with coal to large packages and crates to smoked salmons (a few of which we pilfered and carried into camp at night, tied around one's thigh, which became a bit raw, but the protein was worth it!)

I realize that I mentioned this before but it bears repetition! The death rate at Umeda in our first winter was almost 25% of our group — 115 officers and enlisted men died out of a total of 458! The causes were dysentery, beri-beri, malaria and pneumonia. The basic underlying disease, however, was malnutrition combined with inadequate clothing and practically no heat.

Location of Camp

The Umeda camp was located in an industrial/residential area very near the Umeda Railroad station and freight yards. Officers, as well as enlisted men, worked supposedly for additional rations which never materialized. The real big advantage of working was the chance to increase one's ration by pilferage of calories (sugar) or protein, preferably smoked salmon that required no cooking. After we worked all day, ten hours at least, we came in and ate, talked and tried to reduce our lice population by killing all the adult lice that we could see on our bodies and in our clothes. It was a never-ending task. Our mid-sections were always red and sore from lice bites and scratching. Once in a while, we received a bowl of noodles at lunch. This was really a treat.

We worked six days per week with Sunday to rest, wash clothes, etc. We had very few books, except those we carried with us. Some of us kept diaries. Mine was primarily to do with food and health, sanitation, etc. Since this has been published in summary form in the late 1940's by the Office of Naval Research,

I have attempted not to interfere with the narrative here by flooding it with numbers. Suffice it to say that we were underfed, ill-nourished and cold with many diseases associated therewith. Our stay there, like that at Cabanatuan, can be characterized by existing rather than living.

New Friends made at Osaka Umeda Bunsho

In close quarters such as at Umeda Bunsho, one made friends (or enemies) easily. Among the fast friendships that I made were Dr. David Brown (one of the two physicians in our camp), Major William Massello, Jr. (of Somerville, Mass.), Walter Cadmus, and Col. Theodore Wickord. David Brown was a medical doctor — eye, ear, nose, and throat specialist — who was a battler for the workers, forever arguing with the Japanese about men being too ill to go to work. He had but few drugs to dispense but took care of the troops. We bunked next to each other and when I had a bad attack of malaria, Dave somehow got some quinine and dosed me with an appropriate amount and nursed me through the ordeal. He also helped me get to a Japanese oculist and get new eyeglasses so I could see properly.

Major (Later Colonel) William Massello, Jr.

Bill Massello was a jewel of an individual, an honest, straightforward officer who had been on both Bataan and Corregidor. At Corregidor he literally fired the last big gun (twelve inch mortar) from Corregidor, as the Japanese were landing on the island. He did this while badly wounded, lying on a stretcher; he sent his men under cover from incoming artillery fire, and pulled the lanyard himself. For his heroism on Corregidor he received the Distinguished Service Cross. In Zentsuji, our next port of call after Umeda, he taught a course on the "Great Captains of History." I learned much from Bill and later Diana and I visited them and they us. His wife Olga had been on Corregidor until June, 1941. After he left the Regular Army, he taught mathematics in high school at El Paso, Texas until forced to retire because of age. I still enjoy corresponding with him.

Walt Cadmus became a banker in Oregon and I hear from him on occasion. My acquaintanceship with him was renewed in more recent years through his friendship and business activities with a former student and friend, Marvin Byer.

The most senior prisoner officers at our camp were Lt. Col. G.L. Fields, who died on January 24, 1943, and Lt. Col. Theodore Wickord of the 194th Tank Group. Diana and I met with Ted Wickord and his wife in Chicago after the war. The Wickords came from Maywood, Illinois where his National Guard unit was organized. He passed away some years ago.

We received two meals per day in the camp and one meal on the job. The morning meal and evening meal were basically an inadequate rice ration with a watery soup. Once in a while, in the evening, a small piece of fish was added for each soldier.

Life at Osaka Umeda Bunsho was rough, to say the least, but not as boring as it was at Cabanatuan and earlier at O'Donnell. As I stated earlier, our main evening occupation was not reading but talking. One evening, one of the enlisted men was looking out the third story window near his bunk area when he saw activity of a "coupling" nature in the building close by. He called over his friends and by the time the couple next door were through their sexual frenzy, they had quite a cheering section from the Osaka Umeda Bunsho. Other than this one evening's "impromptu" entertainment, there was none while in Osaka Umeda Bunsho. There simply was no extra space plus the fact that we were just too tired and underfed. The winter of 1942-1943 was the worst of the three we spent in Japan. Shoveling coal, emptying freight cars, carrying lumber, steel, etc., by day and just plain existing the remainder of the time.

CHAPTER 8

ZENTSUJI
(AUGUST 1, 1943 - JUNE 23, 1945)

On August 1st, 1943, forty American officers, including Major Massello and me, left Osaka Umeda Bunsho by foot, train and ferry for Takamatsu, the port on Shikoku nearest to Zentsuji. We proceeded by train to Zentsuji from Takamatsu.

Zentsuji POW Camp was the first POW Camp established in Japan and primarily for the POW's from Guam and Wake Island, as well as some Dutch Naval officers from ships that fought in the battles of Java Sea and Sunda Straight along with the U.S.S. Houston officers, the officers from the British Cruiser, H.M.S. Exeter, and the officers of H.M.A.S. Perth, officers from the Royal Australian Army and Air Force captured at Rabaul, plus assorted other officers and men.

At about the same time or shortly before, some of the officers from the Yodogawa and Tanagawa camps in Osaka were transferred to Zentsuji.

This was not the first time that the Zentsuji Camp was used for POW's. It had been used for Russian POW's from the Russo-Japanese war in 1904 and again in the Sino-Japanese war in 1937. Since the latter date, it was used as a barracks for the Japanese Army.

The Zentsuji POW Camp was located adjacent to a most elaborate Shinto Shrine, and it was by means of recalling this shrine with its three or five tiered pagoda that I was able to locate where the Zentsuji Camp had been when I revisited this area in 1982. This latter visit will be covered in detail later on in this narrative.

The "relatively" well-fed prisoners at Zentsuji (who had not been subjected to the Bataan Death March, nor O'Donnell or Cabanatuan, nor heavy work details in the Philippines), were appalled when they saw us marching into camp — emaciated, lice ridden, dirty, etc.

We were held in a separate "quarantine" area for three weeks after our arrival at Zentsuji. On arrival, we were deloused, took baths and issued clean clothing from the stocks that the POW's, mainly from Guam, had brought with them. We were fed extra rations made possible by the POW's already there getting by with a little less for a few days. The Naval medicos under Lt. Commander Van Peenan and Commander Tilden I. Moe took wonderful care of us. It is something never to be forgotten by me.

General Conditions at Zentsuji

Compared with our previous prison Camps — O'Donnell, Cabanatuan and Umeda — Zentsuji was a "country club;" I use this term carefully and only in a comparative sense.

The food, while insufficient, was far more than we had been issued at the other "hostelries." The officers were coerced into working by the Tojo edict of no work, no food. We mainly "gardened." We broke virgin soil using picks and shovels. I haven't gardened willingly since we left POW status. We also swept the rooms, the compound and cleaned the latrines, plus, on a rotational basis, did all night barracks patrol. The latter was necessary because we were a military "unit" and we had among us still some very sick officers and men who had to get up at night, maneuver down the stairs to the latrines — more than once.

I should say that not all was a bed of roses. We were hazed by the Japanese non-coms, given extra exercise "taiso," etc. The Japanese guards weighed the vegetables we grew and subtracted their weight from the ration we were to get from the Japanese. We also raised rabbits and chickens.

We were fortunate at Zentsuji to have many of the creature comforts we were denied at Osaka, such as blankets and clothing — not as warm as we would have had in the USA in a comparable climate, but an order of magnitude better than at Osaka. Much of the clothing came from the International Red Cross. But there was no heat in the barracks and in the winter time we suffered from the cold, had chilblains. I had to have my wedding ring sawed off with a hack saw because of the swelling of my fingers due to the chilblains.

We had a large 25 ft. by 25 ft. cement bathtub which held 1,500 gallons of water. We were permitted a hot bath and soak in the winter once every two weeks with cold water baths in the summer. The latrines and wash areas were located outside and behind each of the two barracks buildings. Washing with cold water can be quite a "shocking" experience in the winter months and contributed to the development of chilblains on my hands. In spite of this, in spite of the insufficiency of food and heat and the hazing by the Japanese, for the first time in over a year of POW status, we did more than just exist.

To complete the description of the facilities, each of the two barracks buildings, which I am told were formerly warehouses, were two stories and divided into rooms each holding 32 men. The divided space had four windows and two 25-watt bulbs. Sleeping decks raised 16 inches off the floor lined the barracks rooms on either side with two cross passageways in each building. The raised sleeping platforms were covered with tatami mats but no futons. Each of us had 30-inches width of sleeping space. There were door openings to each room but no doors. The food was cooked in a central kitchen by the prisoner cooking detail. For each meal, each room sent someone to the galley for the bucket of rice and the bucket of soup and tea. Each portion was carefully ladled out to be exactly the same as the other. We ate at long tables which were provided, together with benches, in each room. These were also used for reading and playing bridge and chess, etc.

There was also a recreation room and a small barber shop. Bill Massello gave me my haircuts; he was a good barber, among other things; I don't know where he learned how to cut hair, but I was among the beneficiaries of his talents and I had a good head of hair then. There was a PX where we could purchase mainly paper and notebooks, as well as Wakamoto Tablets — a vitamin B complex tablet made by fermentation of rice and then dried. They tasted horrible but I am certain helped alleviate B-complex deficiencies, if one could get them down the gullet, which I did.

Medicine, Surgery and Dentistry at Zentsuji

I became friendly with Dr. Moe who was an internist and who assisted Dr. Van Peenan in his surgical procedures. I believe that during our stay there, with only spinal anesthesia via novocain, Drs. Van Peenan and Moe did 21 major surgical procedures with one of their medical technicians from Guam assisting. A spoonful of sulfanilamide powder was placed in each wound prior to closing. No infections ensued and no men were lost due to surgery by the two wonderful Navy medics. Dental emergencies were handled by Dr. Howard Friedman of San Francisco. He did emergency extractions and, with the aid of a corpsman to operate the GI supplied foot pedal powered drill, did cleaning of cavities and fillings. Usually, he was able to get by with cleaning out the cavities with the various chisels that he had. Dr. Friedman's nickname was "foots;" he had relatively large feet. He was a keen Gilbert and Sullivan fan and I learned many of the Gilbert and Sullivan arias from him. I might add parenthetically, the POW diet or lack of it, and lack of sugar, resulted, in my case at least, in fewer cavities than I have ever had during a comparable length of time.

Dr. Van Peenan was busy all the time in spite of being hungry and cold like the rest of us. He was forever arguing with the Japanese medical doctor, a Lieut. Saito, to keep him away from the patients. By cajoling, persuasion, and simply superior medical knowledge, he was able to get more medical supplies from Saito than was otherwise possible. During the entire stay at Zentsuji — from January, 1942 until we were transferred to Roku Roshi in June, 1945 — we lost only ten men out of a prisoner population of 450 to 950 POW's. This relatively low death rate is attributable to the skill and courage of our medics, Drs. Van Peenan and Moe, and to the medical supplies such as the anti-malarial pills, atabrine and plasmoquine, and anti-dysentery drugs that we received from the International Red Cross via the prisoner exchange ship — the Gripsholm.

One final point that I might make here is that each of us was weighed each month. A measure of health in the eyes of the Japanese? I do not know what they did with the data. I do know that the Commandant, about whom I shall write more later, knew it was in his best interest to keep the prisoners as healthy as possible to get more work out of them, as well as to avoid problems later. I am sure he took each death badly because it could reflect on him at higher headquarters then and postwar, which it did.

Recreation in Zentsuji or How to Maintain Sanity

Books. When the American Embassy in Tokyo was closed at the beginning of the war, Ambassador Grew, prior to leaving Tokyo for the United States, sent his entire personal library that he had in Tokyo, some 200 volumes, to Zentsuji as the nucleus of a library. Later this was supplemented by books from the International YMCA and more from home via the Exchange ship, the Gripsholm. Zentsuji, when we left, had a library of over 3,000 volumes.

I recall taking a calculus book out of the library and doing almost every problem in the book. Most of the POW's read avidly from this splendid library at Zentsuji and I kept a record of the books that I read there. I list those here.

Gino Watkins	*Biography of an Explorer – J.M. Scott*
Alexander Dumas	*Chicot the Jester*
Major Elliot	*Battleshield of the Republic*
	Crosswinds of Empire
Philip Gibbs	*Blood Relations*
	Lord Listeer

	Queen Victoria
	The Enchanted April
Kenneth Roberts	*Arundel*
Francis Biddle	*Mr. Justice Holmes*
Faith Baldwin	*Private Diary*
Victor Heiser, MD	*An American Doctor's Odessey*
Henry F. Pringle	*Biography of Theodore Roosevelt*
Harriet Forbes	*Paul Revere*
Margaret Mitchell	*Gone With the Wind*
Paul De Kruif	*Microbe Hunters*
	Magnificent Obsession
Hervey Allen	*The Forest and the Fort*
Jennings	*Next to Valor*
Pearl Buck	*The Good Earth*
Mary O'Hara	*My Friend Flicka*
Hans Zinsser	*Rats, Lice and History*
	The Autobiography of Henry Adams
Kenneth Roberts	*Captain Caution*
	The Fire and the Wood
	Rabble at Arms
	Northwest Passage
M. D. Story	*Hitler 1933*
A.J. Cronin	*Grand Canary*
W. Somerset Maugham	*Of Human Bondage*
J.S. Van Dyke	*The Streets of Many Rickshaws*
Rafael Sabatini	*The Snare*
	Master at Arms
Uberman	*The Kahn's Progress*
Daphne DuMaurier	*Frenchman's Creek*
W. Delfury	*No Hero This*
Field	*All This and Heaven Too*

Mary Allen Chase	*Mary Peters*
Mary Allen Chase	*Windswept*
Nora Wein	*Reaching for the Star*
Tate	*Stonewall Jackson*
Douglas Reed	*Nemesis*
	Insanity Fair
Lew Wallace	*Ben Hur*
McLaurin	*Post Mortems and Mere Mortals*
Maj. Gen. Sir Frederick Maurice	*Robert E. Lee*
	One Against England
Nordhoff and Hall	*Pitcairn's Island*
	No More Gas
A.J. Hutchinson	*This Freedom*
	The Wolcott Reader
Marguerite Kennedy	*The Fool of the Family*
	The Constant Nymph
John Gunther	*Inside Asia*
	Great Men of Science
	Autobiography of An Adventurer
J.T. Trebitsch	*Lincoln*
J. Scherer	*Three Meiji Leaders– Ito, Nogi, Togo*
Fay	*Franklin*
Walter Edmonds	*Drums along the Mohawk*
Sinclair Lewis	*Arrowsmith*
Richard E. Byrd	*Alone*
Bakeless	*Daniel Boone*
T.E. Lawrence	*Seven Pillars of Wisdom*
Andre Maurois	*Disraeli*
	Ariel
	Byron

Stefan Zweig	*Marie Antoinette*
Oliver Goldsmith	*The Vicar of Wakefield*
Quentin Reynolds	*London Diary*
	Herbert Hoover
Norman Lindsay	*The Curious Amorist*
Englebrecht and Hanighen	*Merchants of Death*
Violet Cambell	*Seed of Adam*
Eric Stanley Gardner	*The Paradine Case*
Pearson and Allen	*The Nine Old Men*
Phoebe Fenwick Gray	*Vivandiere*
Dana Lamb	*Enchanted Vagabonds*
Ambrose Pierce	*In the Midst of Life*
Myron T. Herrick	*Friend of France – Colonel Mott*
John Rathbone Oliver	*Priest or Pagan*

In addition to the above novels, biographies, historical fiction, etc., we had a number of reference books in bacteriology, physiology, forensic medicine, chemistry, etc., which I took out on loan at different times to study, reread or to just plain "cerebrate." We had discussion groups on great literature or books that many felt were "must" reading and which I noted in my diary. These included such works as Eliot's *The Mill on the Floss*, Dumas' *The Three Musketeers*, Bronte's *Wuthering Heights*, Lesis' *Sagitarius Rising*, etc. Today, these are all in the Modern Library series, as well as in other relatively inexpensive sets.

Bridge and Poker; Chess and Checkers. In addition to reading, we played a great deal of bridge. I really learned how to play bridge from some British officers who had been captured on Hongkong or Singapore. I also learned how to play chess, which I had never played before. Our teacher was a white Russian, Mike Ushakoff, who was in Manila and joined the American Army when war began. Mike had been in the Czar's Army before the

Revolution in 1917. Many of my friends also played poker, a game I abstained from. I just am not a gambler.

Discussion Groups and Seminars. I participated in a class given by Bill Massello on “Great Captains of History” and was exposed in detail to Caesar, Hannibal, and Napoleon. I had never really understood much about the military aspects of these three leaders before this course. I am sure that Bill covered more than these three, but these are the ones that I recall. I also studied Russian with Mike Ushakoff. It is amazing what one can learn from good teachers — even without textbooks!

Studies on Malnutrition. In Cabanatuan, I began some studies on malnutrition in POW camps and correlated simple observations with dietary inputs. I extended this work in Zentsuji and measured vitamin A deficiency by constructing a simple biophotometer and measuring the ability of a number of volunteers to see a fixed intensity light source through successively thicker sheets of Japanese toilet tissue. Not a sophisticated test, but at least this got published in *Science* after I returned home. The malnutrition survey was done with the medicos at Cabanatuan and at Zentsuji, but mainly at Cabanatuan where we had a great deal of frank malnutrition. They did there clinical observations, such as wet and dry beri-beri, hypoproteinemia. ariboflavinosis, etc. I recorded the rations and their components — relatively simple at Cabanatuan — as well as the numbers of POW’s being fed. In Zentsuji, I made up five tables summarizing the clinical findings for the month of September, 1943, as well as the protein, fat, caloric, thiamin, riboflavin and vitamin A and vitamin D content of the diets as calculated from the textbooks available in the POW library in Zentsuji as alluded to above. After the war, I received a small (under $1,000) grant to get the calculations refined with up-to-date food value tables and published these in the Office of Naval Research Reports in 1947.

These data showed when malnutrition began for different people with respect to a specific nutrient, and which nutrient became the limiting factor. The extent of malnutrition indicated also the predisposition of an individual to an infectious disease agent.

The Biophotometer studies were done with Prof. Robert S. Harris of the Department of Food Technology, under whom I did my S.M. thesis research and whose name helped me to get the O.N.R. grant. In those days, the calculations were simple but laborious — we used slide rules then!

Quotations. Another hobby became that of keeping a list of famous quotations from my reading. I copied these into a small notebook and up to five or six years ago I still had these notebooks. But somehow, I have misplaced them.

Still another way of "cerebrating" was to recreate some old poems, bawdy ballads and lusty lyrics that two or three of us had learned in college and recreated them from memory. As an example, Bruce Cormack, ex-Cornellian and signal corps lieutenant, my adjacent neighbor on the tatami shelf, and I recreated "Ye Ballad of Chambers St." I still have a copy.

Minstrel Shows and Musicals. On a number of Sunday nights at Zentsuji, various officer groups put on musicals and minstrel shows. We had unexpected talent amongst us. For example, Naval Captain William T. Lineberry, MC, head of the hospital at Guam who was on his last tour of duty at Guam prior to retirement when war broke out, was a remarkable end-man at our American Minstrel Show (Captain Lineberry was actually promoted to Rear Admiral while a POW at Zentsuji). He did not practice much medicine at Zentsuji; he left that for his younger colleagues, Drs. Van Peenan and Moe. Captain Lineberry had a remarkable store of jokes which he told beautifully in his end man role in the minstrel show.

Additionally, we had a group of men from all the services and nations known as "The Bath House Singers" who performed periodically. They sang a number of popular songs (pre-war of course). I always got a lump in my throat and tears in my eyes when they sang "My Diane."

The British officers put on a musical comedy which they called "The Sheriff of Nottingham." One of the hit tunes (and reprises) was:

I am the Sheriff of Nottingham
    I am, Ta-rah-raaah
To the people of Nottingham
    I'm like a father
I have a niece whose name is Jill
    Whose fortune is the reason I'd like to kill
In a pool of slimy, slippery blood,
    etc.

News. Early in our stay in Zentsuji, we received, from time to time, some daily Japanese-English Newspapers, such as the Osaka Mainichi. On August 1, 1944 these ceased. We had to then rely on someone stealing a newspaper while out on a work detail, hiding it on one's person and bringing it back to camp to be translated by one of our two officers who were fluent in written Japanese. From these, plus remarks that a few of the guards made, we knew we were winning the war. We drew maps of Europe and the Far East and updated them according to the Japanese newspaper accounts. Key items such as "The Germans" straightened out or shortened their lines (never retreated, of course) or the naval battles for which the Japanese were claiming "glorious victories" only indicated that America had not only rebuilt or repaired the ships it had lost at Pearl Harbor, but had expanded the naval fleet tremendously. We learned of Guam, Saipan, Tinian, etc., being retaken. Needless to say, our spirits and morale soared.

Radio Broadcast. One day in 1943, I believe it was, we were asked to make a radio broadcast to the folks at home via N.H.K., the official Japanese Radio Network. We talked it over amongst ourselves and agreed to do it providing none of us praised the enemy. It would be a morale booster at home for our parents, wives and children. Many of us participated on August 24, 1943. The broadcast was made in the Commandant's office. It was picked up all over the United States and Alaska by "ham" radio operators on November 29, 1943. Diana received six 78-speed records of my broadcast. When she played it at home, our young dog, a wire haired Terrier heard my voice and began barking. There was no doubt that it was authentic!

Japanese Camp Officials at Zentsuji

During my stay at Zentsuji, the Prison Commandant was Captain Yohei Hosotani, a reserve officer in the Japanese Imperial Army Engineers and a former school teacher. Captain Hosotani could be almost ruthless when he felt something was being done to malign his administration. When I met him after the war, in 1982, as I shall relate herein further on, I learned that he had spent five years in Sugamo Prison after the war for his actions against POW's at Zentsuji.

Captain Hosotani's second in command was a Lieutenant Nakajima whom we inmates nicknamed "Saki Pete;" he was never drunk in our presence, but he always "had a glow on." I recall an occasion when we were aware that there were Red Cross food parcels in the warehouse of the camp and we asked Nakajima when we might be issued some — after all we were hungry; the food had been sent over for us, etc. "Saki Pete" replied with a Japanese proverb: "When the fruit is ripe, it will fall from the tree." We posed the query to him while he was taking evening roll call out of doors and he sneezed. He pulled a handkerchief from a pocket of his britches. As he did so, a package of Lucky Strikes cigarettes fell out of his pocket onto the ground. We decided that we had better speak up before too many more cigarettes and probably food were pilfered.

The third Japanese officer in Zentsuji was the medical officer, Lieut. Saito, who did all he could to thwart Commanders Van Peenan and Moe. In order to stay away from work, one needed to be running a temperature of 102 degrees F. The American medicos usually seemed to be able to get around Saito and keep those with a fever out of work.

Red Cross Supplies

When the exchange ships for the American and Japanese diplomats arrived in Japan in mid 1943, (the Swedish ship, the Gripsholm, was used to return Japanese diplomats), these ships carried food, clothing and medical supplies for the POW's in Japan and other countries in the Far East. The delivery of these supplies was under the auspices of the International Red Cross. This was actually the second attempt of the United States to send relief supplies. The first was in June 1942 and failed because the Japanese would not guarantee safe passage for the Swedish vessel, "The Kanangoora," which was loaded with nearly one million dollars of food clothing and medical supplies.

We learned later that the exchange of diplomats took place at Lourenzo Marquez in Portuguese Mozambique on July 23, 1942. The Japanese were transferred to other ships and departed to Singapore and Yokohama. Similarly, the British and American diplomats at the same time were taken in their own exchange ships to their home ports. On this second and successful exchange attempt, British and Canadian, as well as American supplies were also shipped. The supplies were too late for us in the Philippines, but made a tremendous difference to us in Japan. We received our first medical supplies and food parcels, as well as clothing in Zentsuji.

The Gripsholm made a second exchange trip in December 1943 and carried over not only more Red Cross food supplies but also the greatest morale booster of all — mail! I received plasmoquine and atabrine from Dr. Moe and my malaria was gone forever (I hope). In the Philippines and at Osaka Umeda Bunsho, I had upwards of ten attacks. It's bad business!

At Zentsuji, the record shows that from December 12, 1942 until May 28, 1945, eighteen American Food Parcels were received by each POW plus one Canadian Parcel. In addition each POW received 1/3 of an American, 1/3 British and 1/3 of a Canadian parcel on December 24, 1943.

Certain items were removed such as cheese, corned beef, sardines and sent to the galley to be issued at a later date. The Red Cross Parcels meant the difference between undernutrition, malnutrition and proper nutrition — between life and death. The caloric count of each American Red Cross package varied between 9,000 and 15,000 Calories. They contained usually, one can each of Spam, Klim (powdered whole milk), instant coffee, one high density, 8 oz. chocolate bar, 4 or 5 packs of cigarettes, one deck of playing cards, two or three candy bars such as O'Henry bars, Butterfinger bars, etc., in addition to the items enumerated above which were sent to the galley.

The chocolate bars were really "high octane!" They contained not only sugar, cocoa and cacao bean fat but also the theobromine intrinsic in chocolate. One night, I (and several others) made the mistake of consuming a whole bar instead of one square per day. It was in the middle of winter in the unheated barracks. I threw off the blankets and then could hardly sleep at all due to the theobromine I had consumed in the chocolate bar. Different values were placed on the various items. The common unit of value, used for barter, was a pack of cigarettes. And thus one could trade Klim for chocolate, each having a value in terms of cigarettes. The problem was that many of the officers and men who so desperately needed all the nourishing food that they could get, traded off some of the foodstuffs for more cigarettes. Nicotine is indeed a habit forming weed.

Some of us got together for a major holiday such as Christmas or New years and "quanned" up a bash with a menu, hand done, of course.

One New Years, January 1, 1945, Bruce Cormack, Tom Griffin, Bob Powell, Hank Pierce, Bob Studer and I prepared a meal which consisted of:

Spiced corned beef, with cheese dressing (meat with a Ph.D. degree)
Mocha fruit cake
Demi tasse
Doublemint gum
Cigarettes

The hand drawn menu stated that the meal was dedicated to the great New Year with all its promises, to conviviality, and to the very pleasant pastime of "Bashing the Quan" (from an American/Filipino POW derivation meaning "preparing a food dish or meal").

Much more went on at Zentsuji, such as mail call but that will be treated in a separate chapter. Its importance to a POW cannot be overestimated.

Preparing To Leave Zentsuji

Early in January, 1945, we spotted a small speck way up high in the sky. We heard the guards yell to one another: "B-nijuku; B-nijukuj!" We, the prisoners, were rushed into the barracks. This was the first B-29 bomber ever spotted over Shikoku by our guards much less by us. What a thrill for us!

It was apparent, in June 1945, from the actions of the Japanese Camp officials that something was going to happen at Zentsuji. We were becoming more literate in Japanese as the years wore on and some of the men working in the camp administrative office learned that we were to be moved fairly soon to other camps. Obviously, the Japanese were afraid that there would be an invasion and we, the officer POW's, could be of value as barter or ransom. But we needed to be moved to "higher ground" away from likely invasion sites — hence the move.

Zentsuji to Roku Roshi

We were told by Captain Hosotani on June 23, 1945, that we officers would be transferred to other camps, based on nationality. At 15:00 hours on June 24th, 1945 we lined up in formation and were instructed not to make too much noise and thus "excite" the Japanese civilian population, who, we were informed, did not like us. After the interminable "bungo" (count off) we finally marched off to Zentsuji (electric) Train Station. Many Japanese civilians watched us. If they were angry with us or at us, they certainly concealed it well. (We had heard from our work detail underground that the civilians were anxious for the war to be over, but all of us knew that the Emperor alone could call a halt to hostilities.) We were informed that we could take with us whatever we could carry. Actually, we were lucky to be able to carry ourselves.

We boarded the train for Takamatsu, a bit crowded, to say the least. We arrived at Takamatsu before darkness set in. Again more Japanese civilians were there looking at us apathetically. We crossed to the Island of Honshu, took the train for Okayama where we were put into a large room and given a bento box of rice. All 335 of us remained there for the night — under guard of course. There was no room to move around but we were in better shape than we were aboard the Nagato Maru. Stories abounded such as "We" are graduates of Zentsuji International University with three degrees: BA, MT and DrR – Bachelor of Abstinence; Master of Tolerance; and Doctor of Resourcefulness. The next morning we were fed again, a bento box of rice. Our spirits were high!

After breakfast, we boarded another train for Osaka. The wooden blinds on the trains were drawn by us under orders from the guards. Periodically, along the route from Okayama to Osaka, we peered out. When we had traversed this same route in reverse, in 1943, enroute to Zentsuji from Umeda Station, the whole area on both sides was filled with factories and smoke belching forth. Now it was all in ruins, with an occasional smokestack still standing amidst the rubble. Small wonder that the Japanese guards were instructed not to let us look out the windows.

At Umeda Station we detrained and were transported by trucks to another station which held a train that would take us to Kyoto and Fukui. All around us was devastation at Umeda. The air was filled with acrid fumes and the smell of decaying, burned flesh.

It was obvious to us that the war could not go on much longer. Civilians were living under primitive conditions there. Shikoku had not yet been a military target and certainly had no idea of what was transpiring on the main island. Incendiary bombs had done much damage especially when followed up by strafing fighter planes to keep away the fire apparatus so that the fires could really get going. That night we arrived via Osaka-Kyoto-Fukui-Ono. At Ono, we detrained and the march began, up the side of a mountain (that is now a ski resort) for Roku Roshi (six hills). One of our group of lieutenants was a pilot who had been shot down and with an amputated (above the knee) leg, on crutches, Fred Garret. He, plus another lieutenant with an advanced case of beri-beri, who could hardly walk, marched all the way together. Every step of the six to ten kilometers was uphill. They made it as did we! We arrived about 04:00 hours at Roku Roshi, our last home in Japan for the war. We all cheered when Fred Garrett and the lieutenant with beri-beri arrived shortly thereafter.

At Roku Roshi, we were housed in an old Japanese army summer training camp. The accommodations were dirty, bug-infested, crowded, dark and primitive.

CHAPTER 9

ROKU ROSHI – OUR LAST POW CAMP (JUNE - SEPTEMBER 1945)

When we awakened at reveille the next morning, we observed that we were in a gorgeous, picturesque plateau, more than 6,000 feet above the valley below, in the mountain region of what we later learned was Fukui Prefecture. The Roku Roshi Camp was in a compound of about 225 feet by 450 feet. The perimeter fence was a three split rail fence with guard towers at the four corners. It was easy enough to climb the fence or go through the rails. In fact two POW's did escape later on. But where to?

As pretty as the scenery was, and as nice looking as the outside of the housing was, so horrible was it inside. The main barracks itself was 50 feet by 150 feet, single story. A second floor had been laid on the trusses supporting the roof which area was reached by ladders and was occupied by a number of officers. It was difficult, if not impossible, to sweep the upstairs without getting dirt on those below. A fire pit had been excavated in the dirt floor of the main floor at about the center of the building, but there were no flues for the smoke to go out. The pit was the only source of heat. Lastly, the building was infested with fleas. It had been used by the Japanese Army for training.

One side of the building was lined with single deck sleeping platforms and the other side with double deck sleeping platforms. Twenty two inches of platform space were allotted per individual. There were no futons and, in fact, not even tatami mats. Each POW was provided with five thin blankets. On the approximately 400 gm per man per day ration plus thin watery vegetable soup, I doubt if many could have survived a winter in those facilities.

Feeding Facilities

There was a small kitchen building, the so-called galley, which was very poorly equipped with practically no utensils. Once in a while we received some small portions of fish or meat. All food was eaten in the barracks — there was no mess hall nor were there tables in the rooms as we had at Zentsuji. No mess gear was issued, so we ate from tin cans that we had saved from the previously received Red Cross parcels.

Other Facilities

There was an additional small building, about twenty by forty feet, which contained two large concrete tubs from which we could dip water (cold water). There was only one water faucet for the camp of 390 POW's. Vegetables and clothes alike were washed at the wash rack located behind the main building. The latrine was in a separate building with twelve portioned-off stalls or cubicles with holes cut in the floor for defecation from the squatting position. There was an open urinal at one end of the latrine buildings. We had the impression that the whole facility was such, that regardless of the care we took, it was a facility waiting for a sanitation calamity to happen. Fortunately, we dipped the offal onto the fields on the crops and kept the overflows to a minimum.

Drs. Lineberry and Van Peenan set up a small sick bay at one end of the building. There was a Japanese medical corpsman at Roku Roshi, but we had our own team of physicians. The Japanese, however, gave us no medical supplies at Roku Roshi nor distributed any Red Cross Parcels. We had the medical supplies we brought with us from Zentsuji. In July 1945, we received one of the two standard medical Red Cross chests, and later the second. No surgical chests were released to us.

Clothing

No clothing was received from the Japanese until after their surrender when we received a raincoat, slacks, blouse and shirt. However, clothing was not a problem in the summer, and we had received a sufficient supply at Zentsuji which came from the Red Cross.

Work

Our toil was outdoor breaking virgin soil, planting, weeding, etc. Camp sanitation included emptying out the latrines and spreading the offal on the garden, and road building and wood cutting as added tasks. The officers worked along with the few enlisted men we had. The officers were not required to work on rainy days. However, their ration was cut in half. We were constantly being nagged to work harder with dire threats of cuts in the rations. We wrote a letter of complaint to the Swiss Embassy in Tokyo but the Camp Commandant would not let it go through.

Beatings were the exception rather than the rule here. The Japanese must have sensed that the war was coming to a close. This was exemplified by the fact that the cereal ration grew in July from 508 to 538 grams per day and jumped on August 25th to 661-678 grams per POW daily with a substantial increase in soybeans and green vegetables.

The End Of The War Must be Near

On July 16, 1945, a shuttle of planes (we later heard that this was from Halsey's Carrier Group and also were told that it was the B29's) passed overhead and soon after listening to the "fireworks," one could read a newspaper in the prison compound without any light other than the light from the fires at Fukui about 16 to 25 air miles away. What a conflagration! What really sent our morale soaring was the realization that our naval and army air forces must have run out of worthwhile targets to bomb if they went after Fukui, not even a secondary target, according to Col. Unruh (A.A.F.) the American POW Commanding Officer!

On August 5, 1945, two Army lieutenants went over the wall, or through it. The Japanese "went nuts." All doors and windows were nailed shut. Only the door leading to the latrine was left open. We had to spend a day standing in the sun. Shortly thereafter, the two officers were brought back with their hands tied behind their backs. Later we learned that these two officers were taken to Osaka and imprisoned until the first American Occupying Forces arrived to release them. Thus, they were the first of our group to become truly freemen once again.

A day later, August 6th, 1945, a high ranking officer from Osaka arrived and gave us a speech to the effect that none of us should try to escape because the Japanese civilians had no sympathy for us and were all working to win the war, etc. This was a far cry from Cabanatuan, where escapees were beheaded upon recapture and their heads placed around the perimeter of the camp on spears stuck into the ground. This was in addition to their threats (and action) of punishing by death ten POW's for each escapee, as was done in O'Donnell.

One of our men had purloined a Japanese newspaper in early August and one of our officers translated it in the barracks. It told of a five-power conference at Manila, which Russia tried to dominate, according to the newspaper, but the United States did not permit it. Thus, we had another clue of the end of the war approaching.

On August 16th, 1945, the Japanese guards became overly vindictive. We rationalized that the Japanese must have suffered a major setback and the civilians had just been made aware of it. We protested to the Camp Commandant, but no reply was received.

On August 17th, 1945, the Camp Commandant was informed, by Lt. Col. Fitzpatrick that we were quitting work, "period," "by order of the Committee of Senior Officers" (yes, we had committees even in prison camp; this is not a modern day phenomenon at all). In the same communication Col. Fitzpatrick requested an immediate increase in foods — rice, beans, potatoes, etc., and that officers and enlisted men receive the same ration. We were then informed by the Japanese that there would be no further work by the prisoners until further orders. The air was tense. What was going on? From one of the guards, we learned that the Camp Commandant had been called to Osaka to a meeting. We knew that the single communication to the camp was a one-line telephone into the commandant's office and that he had received a telephone call on the morning of the 17th. We heard the telephone ring! It was a small quiet place.

On the morning of August 21, 1945, we were awakened by an American Reveille. One of the POW's borrowed the bugle from his Japanese Guard counterpart and blew a "proper" American Reveille with Japanese guards standing nearby, most affable.

August 22, 1945 Announcement of War's Ending; Raising the Stars and Stripes over Camp Mollette

On August 22, 1945, the Camp Commandant returned and sent a message to our barracks that he wanted to meet with our senior officers — majors and above. He came to our barracks and took off his uniform cap. He had never done that before. He met with the senior officers in one of the smaller rooms and he stated that "the Emperor" had brought peace to the world; that "Japan and America were no longer at war." Large "PW" signs should be painted on the roof. He further stated that he and his group would remain with us for our "protection."

We junior officers had been eavesdropping, but remained quiet until the Commandant left. Colonel Unruh, with his hands shaking, came out to make the announcement. It was 16:30 hours on August 22, 1945. "The War is over." "We don't know yet when we will leave Roku Roshi." We immediately had a flag-raising ceremony. Two American flags had been smuggled from Bataan and hidden throughout the war and its Death March to that day! How they ever were kept hidden from the Japanese through the various, innumerable body searches, surprise inspections, Lord only knows. But down came the Japanese flag and up went the American Stars and Stripes. We were all crying as we saw the flag being raised, saluted, and sang, "America," and "Praise God From Whom All Blessings Flow." All the Japanese guards and Lieutenant Habe were ordered to give "Eyes Right" to "Old Glory."

FIG. 9.1. READY TO RAISE THE STARS AND STRIPES AT ROKU ROSHI POW CAMP, AUGUST 22, 1945
The U.S. flag had been carried by a POW since the surrender at Bataan. The camp buildings are in the background.

We renamed Roku Roshi Camp as "Camp James J. Mollette" after the officer who died there. His gravesite was pointed out to us when Diana and I visited the campsite in the early 1980's, as I shall relate further on. We took down the fences and guard towers. We were free! We wandered into the village. Two of the enlisted navy men wandered far and purloined a radio which was in the Mayor's office and we got a bit of news — in Japanese. They also managed to find "some women of easy virtue" either in Roku Roshi or Ono, and came back to Dr. Lineberry, demanding a "prophylaxis." He like others of us wondered where they got the strength! Captain Lineberry, then in his sixties must have felt like

Longfellow …"The thoughts of youth are long, long thoughts"... I might add that the two enlisted men left the Mayor of Roku Roshi a receipt for the radio!

Certain rules were established by the now ex-POW's of Roku Roshi.

First:	Everyone would stand roll call at both reveille and retreat.
Second:	Japanese guards and camp officials would not be harassed by any of us.
Third:	Japanese civilians would not be molested; our conduct must be exemplary as that expected of an American officer and gentleman.

Needless to say, these rules were strictly adhered to at Camp James J. Mollette in Roku Roshi, near Ono, Japan until we left.

August 24, 1945 More Food To Come

Announcement was made of comfort kits to come the next day, as well as meat, fresh fruits, sugar, dried fish and an increase in the rice ration.

September 1, 1945

Colonel Unruh, Captain Lineberry, Lts. Reder and Johnson returned to Camp James J. Mollette with two tons of relief supplies. Meanwhile, the Japanese had plied us with 17 pounds of green grapes per man!

September 2, 1945

A navy plane buzzed us. We were all excited. He hovered around until another Navy plane came over and "bombed us" with a parachute with a large cylindrical container containing a crate of eggs. Then the “bombing” began in earnest with five B-29’s which buzzed the camp and then made a drop. From their cavernous bellies, 100 parachutes of food, clothing and medical supplies came down! Then they flew over us again at 200 feet or so of altitude. What a sight! These were the first B-29’s we had seen except for the high flying one several months earlier over Zentsuji.

The radio helped bring us news of the war. Russia had entered the war; we had dropped a new weapon on Hiroshima and Nagasaki.

The Atomic Bomb

Should we have dropped the bomb? Naturally I am biased on this matter. The atomic bomb shortened the war. We could never have withstood another year of captivity under the primitive conditions of Roku Roshi and the lack of food, lack of heat, lack of woolen clothing and of medical supplies. Moreover, the Japanese Army would never have given up without a major battle on Japanese soil with countless casualties on both sides. The atomic bomb brought rational thinking into play and convinced the Emperor.

The official surrender of Japan, as we know, took place on September 2nd, 1945 aboard the USS Missouri. We learned later, that while our camp was known to American authorities, the pilots had trouble locating it — hidden as it was in the Japan Alps — until finally a Navy pilot spotted the PW on the roof of this small barracks building. From September 2nd until we left, we gorged ourselves on our new riches of food and more food, pipe tobacco for those who wanted it, cigarettes, clothing for those who needed it (all of us), and medicines.

On September 8th, a recovery team of U.S. Army soldiers, medics and nurses arrived. We were so interested in the new battle uniforms and carbines, etc., that the nurses were hardly noticed until one of them spoke up and said something to the effect of “how about us.”

CHAPTER 10

THE ROAD HOME
(SEPTEMBER 9 - OCTOBER 30, 1945)

Roku Roshi (Camp James Molette) to Fukui

On September 9, 1945, we left for Fukui by truck via Ono with the Recovery Team which, I should have noted earlier, consisted of a squad or so of soldiers from the First Cavalry Division, two doctors and two nurses, and an interpreter (a Turkish citizen with the Turkish Consulate in Kobe). Our senior officers, two days earlier, had decided that the B-29 drop of food rations was beginning to run low and so they sent one of our POW majors, Major William Orr, and another officer on September 7th, 1945, to Kyoto and Osaka to liase with the American troops there. They returned early the next morning, on September 8th, to inform Col. Unruh that the recovery team would arrive later that day, September 8th, which they did at 09:00 AM in a convoy of Japanese trucks. We could see the convoy snaking its way up the mountain road slowly. We left Roku Roshi the next morning, September 9, 1945, by truck, for Ono, down the mountain (6,800 feet up). At Ono we switched to electric trolley trains for Fukui.

At Fukui, we saw the tremendous havoc raised by the fire and explosive bombing of July 16th which lit up the skies as far as Roku Roshi. We were given tea and Japanese hardtack. We were served at tables and chairs by ladies from the Japanese Red Cross.

Fukui to Yokohama

At Fukui, we boarded a special train for Kyoto and Yokohama — with plenty of room for all. We could see tremendous devastation all along the route, especially between Kobe and Osaka, which once had been one building after another, for several kilometers, housing aircraft factories — all leveled.

We arrived at Yokohama on September 10, 1945 at 7:29 AM. We were met at the station platform by Lt. General Robert

Eichelberger and General MacArthur himself, together with the First Cavalry Division Band playing "California Here I Come." It was a thrilling event. Armed GI soldiers were guarding all approaches. They had the look and appearance to us of young gods — healthy!

From the RR Station at Yokohama, we went aboard more trucks bound for the port area of Yokohama. There we were fed an American breakfast of hotcakes and real syrup. Then we were all run through a hot bath, at the other end, our clothing and possessions were deloused, and we were all issued new clothing. A group of medicos and nurses were making hasty decisions as to whether each individual was fit to travel. Rest assured we were all ready, able and more than willing.

Effective September 2, 1945, each American Army Officer was advanced one grade in rank. This was from General Order No. AG-AD-77 of MacArthur dated September 16, 1945. The U.S. Navy officers fared better; they advanced with their class. Thus, they went up in rank three and even four steps higher.

Yokohama to Manila Aboard USS Tryon

This ship was an attack/hospital ship. It was a troop carrier designed for invasions — both for transporting troops in and for handling casualties after an invasion. As I boarded the Tryon, I heard my name yelled out from the ship and was hailed by one of the Tryon's naval lieutenants. I looked up the gangplank and saw that it was an old chum of mine from the Alexander B. Bruce Grammar School on Ames St. in Lawrence which I had attended from 1928-1932. I had not seen Tom Lardner since 1932 — some thirteen years earlier. And he recognized me! He had heard that I was a POW and was wondering if I would show up on his ship. Tom had grown up on Margin Street in Lawrence. He recognized me despite my thin appearance. I was thrilled to have been recognized by someone from my hometown, 9,000 miles or so away, and he was equally thrilled to be the first local to greet me on release from captivity.

After we had boarded and were shown where we were to be berthed, we were informed that the senior officers, major and above only, due to lack of facilities' space, would have use of the ship's officers' wardroom facilities. Needless to say, when Tom heard that, he made arrangements with the skipper of the Tryon for me to use the Ships Officers' Wardroom as Tom's guest!

As a matter of interest, at Roku Roshi, I had my measurements taken on August 26, 1945. They were:

Weight	134.2 pounds	(Normally 165 to 180 pounds)
Neck	13.75 inches	" 15.5 to 16 inches
Chest	33 inches	" 42 to 44 inches
Waist	28.5 inches	" 34 to 38 inches
Hips	36.5 inches	
Wrist	6 inches	

Thus, I was surprised that Tom Lardner recognized me. Most of us, other than the field officers and a few ranking army captains, were fed cafeteria style. But no matter; ice cream, etc., was served three times per day! There were a total of 5 Lt. Colonels, 34 majors, 59 captains, 98 first lieutenants, 25 second lieutenants, 7 warrant officers, and 11 civilians aboard to be berthed and fed three square meals per day. I have no idea of the numbers of enlisted men being taken home. I went to sick-call once due to a "jittery" stomach and recall being given some Tincture of Belladonna, a drug I was to get to know well later in the 1960's when I developed duodenal ulcers.

Manila, September 15, 1945 to September 30, 1945

We arrived in Manila on September 15th, 1945, and were taken to the 29th Replacement Depot. The Depot was activated on July 1, 1945, with a primary mission to forward replacements to the fighting units from the United States. Its primary mission, after the surrender of Bataan, was to return ex-POW's to the United

States in as good a condition and as quickly as possible. It was located on the Manila-Sanya Rosa Highway, eighteen miles south of Manila.

We were to be given complete medical checkups, vaccinations, typhoid and tetanus shots, etc. In addition, we were given some partial pay and uniforms. Hair cuts on the base were free! Supposedly, they would send cables, at our own instigation and direction, to our next of kin. I wrote several but Diana never got any of them. There were many forms to fill out — for service records, pay, etc., as well as forms dealing with atrocities that had been committed upon us or that we had observed. The time moved rapidly for the first few days and then it was really slow waiting for a ship to come in, and then to be off-loaded, loaded and back on the long road home.

While returning to my tent from an outdoor shower, one day, my hand must have been very slippery and I lost my wedding ring which had been cut in Zentsuji due to chilblains. I felt terrible about this inasmuch as the ring had been handmade for me by Diana's Uncle David Goldwasser and given to me on our wedding day by Diana. I retraced my steps but to no avail. Moreover, I learned later that Uncle Dave died in October 1945.

I recall visiting Oscar B. Mapua, my dear classmate, during my stay at the Depot, visiting the main shopping area on the Escolta. I visited the Philippine-American Drug Store in which I had last been on December 31, 1941, and just spent the time "lazing around." I was beginning to put on weight and feel strong once again. We were not encouraged to leave the base overnight but could get special passes if we wanted to do so. I saw no reason to do so.

Manila to San Francisco, October 1 - October 22, 1945

I was in a group of 111 recovered officers and enlisted men who boarded the USS Bolivar on September 30, 1945 and weighed anchor on October 1. We ranged from Corporal to Full Colonel. I note that we were all from New England, New York, New Jersey or Pennsylvania. It was a smooth trip with nothing much to report on. I should state that throughout my trip back from captivity, I kept on trying to send cables to Diana but she got perhaps one or two.

Along that line, we (the Army personnel) took a long time to return. The Naval Officers and men were given the option of returning by ship or flying. One of my Zentsuji friends, Frank Wolfsheimer, who was captured on Guam, returned a month before I did, and called Diana to tell her I was all right and that I was enroute home, albeit more slowly. Frank was asked what route he would like to fly back and he asked if he could return via Guam where he was captured. Of course, the Navy willingly took care of his request.

Passing under the Golden Gate Bridge, October 22, 1945

On the 22nd of October, 1945, we passed under the Golden Gate Bridge. What a thrill! In Zentsuji, we had a bunch of rhyming verses such as Golden Gate in forty-eight, or home alive in forty-five, etc. We were put up for a few days at the Presidio as outpatients of the Letterman General Hospital. There, more medical checkups, more inoculations, a chance to get some more pay, purchase some gifts for Diana, my mother and her mother and await the makeup of the next Pullman train traveling east.

Diana wanted to meet me at the boat but was dissuaded by well-meaning friends. She was told that there were no hotel rooms, travel was difficult, etc. So, it was two or so weeks longer. I visited the Habers who had been so nice to me on Yom Kippur, 1941, and they were surprised to see me inasmuch as they had heard that I had not made it. As soon as I could get free to get to a telephone, I called home and spoke with Diana, her parents, my parents, etc. It was just great. The bill I ran up was big by those standards. I hate to think what it would be today. But we became closer at a distance.

San Francisco to Fort Devens, Ayer, Mass., October 24-30, 1945

We traveled via special train of Pullman Cars as soon as they had a sufficient group to fill the cars. For Fort Devens, enroute to Lowell General Hospital we had two officers and 32 enlisted men. When we arrived in Troy, N.Y., the evening of October 29th, I looked for a place to make a telephone call to Diana and to inform her that I would be arriving at Ayer the next morning. We were parked in the railroad yard and the train was being refueled, oiled, watered, etc. On asking around, I found a telephone not too distant — at a local "house of easy virtue." The madam was most accommodating and allowed me to use the telephone to make a collect call. I reached Diana and she told me that they had, through friends in the Red Cross, been following our progress east and that we were due in at seven the next morning. She knew more about our ETA than we did.

Indeed, we did arrive at Fort Devens about 7 AM on October 30, 1945, and there was Diana, our three and one-half year old son, Errol, whom I had never seen nor who had ever seen me, my parents and Diana's parents. Errol knew I was daddy but really didn't know me. We slowly but steadily became acquainted. My father hugged me and "benched Gomel" (thanksgiving on returning from a perilous journey). Indeed, it was that and something really for which to be grateful to the Almighty.

I was processed later that morning at Fort Devens and immediately released on 30-day leave which was later extended to 120 days.

Early Days of Freedom

During the 120-day terminal leave, we began to plan. To us both it was obvious that I needed graduate education, if I wanted to really get anywhere. But first, we had to really celebrate Thanksgiving. Diana's mother cooked up a storm and this was quite a contrast to Thanksgiving in Moji-Osaka, 1942.

We visited with friends and lounged around; I periodically left the house for a walk. Diana's mother could not understand why. But God rest Diana's soul, she did. She told her mother that I was just tasting freedom again.

The local Temple which Diana's parents co-founded had me speak one evening, and I spoke at a Sunday morning breakfast at the local YMHA on Concord Street in Lawrence. I visited MIT and thanked Prof. Horwood for writing to me in prison camp and picked up some application papers for admission to Graduate School in Biology to major in nutrition. I was accepted.

FIG. 10.1. DIANA AND I IN MIAMI BEACH IN FEBRUARY 1946
(Courtesy of U.S. Army Air Forces)

The U.S. Army History Branch made me an offer to become a field officer, if I would join the Regular Army and assist in writing up the history of the U.S. Army in the War. I thought it over, and talked it over with Diana and decided that I had been away from family and friends for too long. I enjoyed freedom and lack of regimentation. Moreover, I needed more education, and we then had the G.I. Bill of Rights! Diana, bless her heart, had lived with her parents and saved most of my pay allotment which came to her, so we had some cash in the bank to supplement the G.I. Bill. Plus the fact that I was sure that I could get a scholarship or financial help of some kind.

In February 1946, the Army Air Forces, of which the 803rd Engineers was part, gave us a first class train ticket (Pullman) to Florida including a two-week stay at a first class hotel as part of my recuperation. Diana and I had a ball! Both her Uncles Max and Sidney Greenberg and their wives were in Miami at the time and we really ate up the sun and the good food. Life was good.

We bought a second-hand car and so could travel around easily — when it worked. It wasn't such a good deal. Most second-hand cars are the previous owner's problems, and so it was with this one.

Before closing my abbreviated military career, however, there are three more areas to be covered in the next chapter. These relate to certain matters and specific people without whom I doubt if I would have survived the POW ordeal.

CHAPTER 11

SURVIVAL

The question as been asked of me frequently, especially for the decade after the war, "How did you survive it, mentally, physically emotionally?" The quickest and easiest answer is "with a lot of luck and the help of God." One thing is certain and that is that there were no atheists in the foxholes of Bataan. This answer ill-satisfies many, even most, but not me nor any thinking person. I discuss these in what follows, in random order. No single factor is more important than any other. All of them were of vital importance at different times in my own case.

Good Physical Condition

At the time of surrender, this was very important and continued to be so throughout captivity. The best proof is that in the Cabanatuan POW Camp, the death rate of those who were captured on Corregidor was far less than that of the Bataan survivors who had suffered the ordeal of The Bataan Death March, and a month longer on poor food with more malaria and dysentery among us from Bataan.

Genetics and The Will to Live (Motivation)

Obviously genetic makeup is of importance. We are just now really beginning to learn "how important it is to pick one's own parents." Of similar importance and probably connected thereto is "motivation." It was just too easy to plain give up. Captain Herb Ingersoll, God rest his soul, gave me the assist I needed on the Death March at one point. Human endurance just went so far. Without Herb "goosing me on," I doubt if I would have made it.

Luck

Extremely important. But the old adage of "The harder I work, the luckier I become" applies also. I took a hand in getting lucky by chancing to write a note to Oscar Mapua, my classmate at M.I.T. I was lucky that the caribao driver delivered it and came back with the money which enabled me to purchase drugs — mainly quinine. This kept me alive until Osaka when Dave Brown gave me more quinine when I had a malarial attack. Another example is going on the Mountain Province work detail and meeting the Nuns of the Order of St. Augustine in Bontoc and Lubuagen, and especially Sr. Jean Delbecke (formerly Sr. Marie Arthur).

Belief in a Supreme Being and Prayer

Religious belief, I found was extremely important. I do not recall the specific religious denomination of the Protestant Padre whose services I attended once in a while at Cabanatuan and at Zentsuji, nor do I recall their names, but I still remember some of the hymns that we sang and the pleasure that they gave me and I suspect, the spiritual uplift. I did not have a Jewish Holiday calendar in the camps, so I did not celebrate the various Jewish Festivals, but I thought about them and the meaning they had for me. I appreciated more than ever, the spiritual significance of the Passover Festival revolving around freedom. Belief in the one God took on much meaning to me in the various camps. I had someone to talk with silently, to pray to for help when I needed it.

Education, "Sechel," Native Intelligence

Education in my case was vital to my survival. Prof. Murray P. Horwood wrote me in Zentsuji saying that he did not know where I was — the Philippines, Japan or wherever — but that he hoped that the subjects of microbiology and sanitation that he had taught me would serve me in good stead. They did indeed, Prof. Horwood, as I informed him when I returned.

"Sechel," native intelligence or an uncommon amount of common sense, went a long way in helping us to survive. Intelligence in dealing with other people played a vital role. "No man is an island unto himself." How true that is in POW camps.

A friend of mine, a superb microbiologist, Dr. Ralph G.H. Siu, is a deep thinker and philosopher and has written a number of books including "The Tao of Science" — his spare time work. Ralph has stated that "Philosophy should be taught in the university by a group of people, all over 65, from diverse backgrounds, because anyone who has lived that long has been successful and must have a philosophy that worked for him."

Native intelligence enables us to accept, adopt or modify what we observe or experience and helps in survival. Our education can aid native intelligence but not replace it.

Mail: The Big Morale Booster

One of the big worries of a POW is his family and how they are faring. He knows how he is but they probably do not and are concerned about him. Did my family know I was in the Bataan Campaign? That I survived? That I was captured and in various POW Camps some of which were hell. If the answer is in the affirmative, then they must have been worried terribly. All of these questions passed through my mind at night in the various POW camps until the mails finally came through. We did not know, for example, whether the Japanese gave information to our loved ones about whom they held as POW's.

I knew that Diana was pregnant when I went overseas. I learned later, after I returned how her father had tried, through the good office of his friend, Congresswoman Edith Nourse Rogers, to get word to me that Diana had been delivered safely of a healthy baby boy while I was on Bataan. But, because of the emergencies of war, personal messages were not accepted. However, she received two cables from me dated April 2, 1942, sent from Cebu, while I was on the front lines of Bataan, that I was okay and feeling fine. So she knew that, as of April 2, 1942, I was alive.

Diana heard from the War Department, after the fall of Bataan, that due to the "nature of the enemy," I would be kept on the rolls as "Missing in Action." At least she would be getting her allotment from my pay of $125.00 per month. Indeed, I was so regarded by the War Department from April 9, 1942 until August 25, 1943, when the kind postman asked her when she had last heard from me or about me. She told him and he then proceeded to tell her that I was all right and not to scream or shout as he handed her a card from me from Osaka Umeda Bunsho, which, of course, she began to do and her mother rushed out to the front door to see what was going on. My Missing-in-Action days were now over. Diana notified the War Department and was informed by a letter from Howard F. Bresee, Colonel C.M.P., Assistant Director of the Prisoner of War Division of the Office of the Provost Marshal General in Washington to whom she sent a photostatic copy of my postal card via the American Red Cross, that:

> ..."The prisoner of war postal card submitted by the American Red Cross will be accepted for placing his name (SAG's) on the prisoner of war list."

The letter then went on to state that my card was being referred to the Adjutant General for further recording and verification and that when negotiations were concluded for a third exchange ship at some future date, parcel labels would be sent to the next of kin of POW's listed on the records of the office. They then enclosed instructions as to how to communicate with me.

On the same day that Diana received my card, she proceeded to write me again (having written me a number of letters earlier-on that had been returned to her). In her letter to me of August 25th, 1943, one page, handwritten on both sides, was the news that she had had a 6 pound 14 oz baby boy after 6 1/4 hours of labor. She also enclosed a picture of her and Errol David together with our late sister-in-law, Leah Greenberg. In the letter she also described our son and his activities. The first letter was received by me on March 23, 1944 at Zentsuji. I received her second letter to me, written a day later, (August 26, 1943), on March 27, 1944. I had been more than two years without mail from home and it was almost as long since she had heard from me. Meanwhile, of course, the stories of the Bataan Death March and other POW atrocities had became known to all. Well, at least she knew that I was alive about six months earlier.

In her letter of August 26th, she described the many activities she had undertaken on behalf of the war effort — knitting sweaters for the American Red Cross, USO work and that she had begun to take a nutrition course through the Red Cross and was doing volunteer work at one of Lowell's largest hospitals as a Dietitian's Aide. She informed me that most of the officers' wives of the 803rd Engineers were writing letters and sharing any bits of information that they had received. It was through a letter received by one of my fellow officers from his wife that I first learned of Errol's birth.

I was reminded, in a letter dated December 21, 1945, from a fellow POW who had been on a Norwegian Freighter that had been sunk off Cebu and had joined the U.S. Army as a second lieutenant, that on Errol's birthday I used to hand out cigarettes which I had saved from Red Cross Parcels.

In 1943, Diana had also received two letters from me mailed from Bataan on January 4 and January 16, 1942. They must have sent the mail out by submarine, because there were no planes or ships coming through at that time. We were all cut off by the Japanese.

The Japanese instituted, some time in 1944, I believe, 25-word maximum-length letters presumably to save on censorship efforts. All letters, except the first ones that I received were censored both by the United States and by the Japanese. Thus mail was at least six months old when we or our intended addressees received the letters in the states or in Zentsuji. I received a number of 25-word letters from my parents and a few from others than those mentioned above. Everyone wanted to be sure that if there was a limitation put on the numbers of letters that I could receive, that Diana's letters were not cut out. I still have all of the letters I received in prison camp with the date I received them handwritten on each one.

Letters from home meant much to us. To some of us, mail day was more important, in a way, than Red Cross Parcel issues. One could eat the food only once, but one could read and reread the letters. Mail from loved ones was the greatest morale booster of all for most of us. Knowing the news from home, and to know that all was well, made it possible for us to withstand further privations. We could look forward, Deo Volente, to a bright future. An unfortunate few of my fellow inmates received bad news, such as deaths of loved ones, separations or divorce. But, the latter two were relatively few.

I suspect that the reason I saved the letters and cards that I received was because of their importance in maintaining my emotional well-being, and a reminder of how much my young wife, and indeed my entire family, meant to me.

Red Cross Food Parcels, Medical Supplies and Clothing: The Difference between Life and Death in POW Camps

As mentioned earlier, the diplomat exchange ships also brought desperately-needed medical and pharmaceutical supplies, as well as hand knit olive drab wool sweaters, books and the food parcels.

The drugs for curing malaria, Atabrine and Plasmoquine, were new to us and saved countless lives in prison camps. The pharmaceutical supplies also included sulfa drugs for bacillary dysentery and drugs for other tropical diseases. They also sent plenty of vitamin concentrates and these, together with the food parcels, alleviated the diseases of malnutrition. We certainly never expected to become fat again or approach our normal weight through consumption of these supplements, but we would not so soon become emaciated as we were. The chocolate bars and sugar especially helped make life more bearable.

I am convinced that the medical/pharmaceutical supplies, as well as the food parcels were among the key factors in keeping the deaths at Zentsuji as low as ten men over a four-year period. The other factor was of course the wonderful medical care we received from Drs. Van Peenan and Moe.

Raisins and prunes were included in the Red Cross parcels. They were sources of instant energy. Canned butter, and citrus concentrate helped to provide some gastronomic and gustatory delight.

Angels of Mercy and Later Good Friends for Life — the Belgian Order of the Sisters of St. Augustine — Especially Sr. Jean Delbecke (Sr. Marie Arthur)

For the relatively few of us who were in the Mountain Province work detail, these wonderful nuns helped to save our lives by food, drugs, nursing and human kindness. I have discussed earlier their help to us on the Mountain Province work detail in May-August, 1942. Not only did they give us drugs and medical supplies, but they bandaged our feet and wounds — often at risk to their own lives.

I can still see in my mind's eye, Sr. M. Arthur (Jean Delbecke) rushing along the tops of the rice paddies, holding up the hem of her skirt with one hand and lugging a basket of food, drugs, etc., with the other. And leading a small troop of her young school pupils who were laden with freshly prepared foods for us, and who were housed in the evening at the small school house known as Pud Pud. They were kind and brave beyond measure. I had the pleasure of visiting them in the Philippines several times in the 1960's and 1970's.

In 1961, when I knew I would be attending the First Japan Atomic Energy Forum, through former students, including post doctorals, from the Philippines, I became acquainted with Rosalinda Valenzuela, M.D. (aunt of Maria Clara Valenzuela of the I.L.P.). Dr. Valenzuela worked for the Philippine Department of Health and she located Sr. M. Susannah at Baguio and Sr. M. Arthur then moved from Lubuagen to St. Theresa's College at Quezon City.

Diana and I met Dr. Valenzuela in December 1961 in Manila, after the Atomic Energy Conference in Japan was concluded. She went with us to Baguio, where we met Sr. M. Susannah and her Mother Superior, and then to Quezon City for a joyful reunion with Sr. Marie Arthur. It was a pleasure for me to thank them in person and to introduce Diana to them.

Sr. M. Arthur, our Lubuagen savior, had charisma, beauty, kindness and love in her and written all over her face. She and I corresponded frequently and I saw her twice more in the Philippines and, in 1964 had the pleasure of bringing Jonathan and Diana to introduce the "baby" of the family to Sister. She kept a picture of our family under the glass top of her desk. When I got a chance, I spent the weekend of a trip to Japan, in Manila and visited her.

I had asked Cardinal Cushing what I might bring or send this lovely lady. He advised, "gifts get lost in transit." He had much experience in this in South America, but a nun always had use for a few dollars for her favorite charity or her school children for clothes or books. So I brought money and gave it to her when I was there or through friends, if they were going there.

After Pope John XXIII, of blessed memory, liberalized the Church and the nuns began to wear regular clothes, Diana and I sent her a small portable silver altar from Rome and a pearl crucifix from Bethlehem in the early 1970's. One time when I visited her, I brought her a simple watch from Japan which Masa Ayukawa helped me pick out so as not to be too ostentatious, and a silver M.I.T. Cross pen. She was using them until she passed away, and I noted with pleasure that she was wearing the watch when I visited her at her sisters' home in Rouleraat, Belgium in the early 1980's. All of this gave me much pleasure. She was like one of our own family.

When she suffered a stroke in 1979, through friends like George Sankey and Justin Bloom of the American Embassy in Tokyo, I was able to get the American physician at the Manila U.S. Embassy to call on her and he was prepared to do all he could to help should she need any. It turns out that she did not. I had pointed out to the Secretary of State of the U.S.A., Mr. Cyrus Vance, how she had saved so many American lives at risk of her own during World War II and should be treated as we would one of our own Army/Navy nurses. Indeed, our Embassy at Manila were ready to serve as needed.

After he visited Sr. Jean at St. Theresa's in Quezon City, Dr. John H. Baker, the Regional Medical Officer, wrote me:

> …"It was a pleasure for me to sit for a time with these gentle ladies and reflect upon the qualities of a faith which in a cruel time made their responses so humane in a period of inhumanity. I was reminded of the closing lines of George Eliot's novel, "Middlemarch," ..."the good of the world is partly dependent on unhistoric acts, and that things are not so ill with you and me as they might have been is half owing to the number of those who live faithfully a hidden life and rest in unvisited tombs"… I felt it a unique privilege to call upon Sister Jeanne"…

Sister Jean Delbecke moved to Belgium in 1980 and I visited her at the home of her two sisters, in June 1980, at Roesaelare. She was retired then. It was the last time I was to see her alive. She died on September 27, 1982 at Heverlee (near Louvain where the Order was headquartered and where they had a home for retired nuns). She died within a few hours of her brother Joseph.

I visited her grave with Sr. Lieve, the head nurse at the Pavilion at Heverlee in early 1983. My wife and our entire family lost a wonderful friend who was my rescuer when I was in need. The warmth of her personality and the kindness of her soul will ever be green in my memory till time shall be for me no more.

CHAPTER 12

GRADUATE STUDIES (MARCH 1946 - FEBRUARY 1949)

…"Lives of men and women are often touched and shaped by events and experiences as well as by people with whom one comes into contact"…

Applying to Graduate School at MIT, In Early 1946

In addition to the Regular Army promotion that I had been offered, I also ascertained that I was welcome back at Arthur D. Little, where I had worked before going on active duty. While the immediate financial rewards were less, there was no doubt that graduate studies were necessary. The area I was looking at, nutrition and foods, seemed to have plenty of needs for students and fellowships/research assistantships. Moreover, the G.I. Bill of Rights would really help, in addition, with $90.00 per month plus books, supplies and tuition!

On December 4, 1945 I applied for admission to Graduate School at M.I.T. to study for the S.M. degree in Nutritional Biochemistry. While I had but mediocre grades as an undergraduate, in my application for admission, I cited the work that I did in POW camps. In the event, I was admitted for the S.M. degree in the Department of Biology. Prof. Robert S. Harris, under whom I had done my S.B. Thesis, offered me a research assistantship in nutritional biochemistry with a stipend of $100.00 per month. Between that and the G.I. Bill, plus the savings Diana had accumulated during the war from the allotment of $125.00 per month that she received from my pay and allowances and my own accumulated back pay, we should be able to make it. I did not apply for the doctorate program at that time primarily because I was not sure that I had the necessary "Sitzfleish" for the three or four year program. And, I had been away from studying for over five years.

On one of my early visits to M.I.T. after I had returned, I ran into Prof. Bernard E. Proctor, the senior faculty member in food technology, at the Smith House Restaurant located adjacent to the Burton House Apartment complex. He had heard that I had "made it" and was okay and asked me to come by and see him whenever I was at M.I.T. again. I dropped into his office shortly thereafter and he suggested that I might like to come and study Food Technology in the new department that Prof. William L. Campbell and he were forming.

I informed him that I had just accepted a research assistantship under Prof. Harris in nutritional biochemistry but that should I do well enough and a position was available for doctorate work, I would like to ask him for consideration. Prof. Proctor left the door open for me to talk with him further whenever I so wished.

In the meantime, I planned to commute by car or train to the Institute. We were still living with Diana's parents in Lowell — rent- and board-free. They refused to take any money then nor for the four years that Diana and Errol had lived with them during the war. In effect, we were subsidized by them and hence the savings account. They were wonderful people and most considerate.

Shortly after the war ended, M.I.T. had set up some new graduate housing for returning married veterans, located west of the playing/drill fields using prefabricated house blocks. I applied for one of these units soon after I had received a letter of admissions to Graduate School from B.A. Thresher, M.I.T. Director of Admissions, because I was spending close to three hours per day commuting to and from Lowell (recall that this was before the days of superhighways. Three lane roads for both directions was the norm for most intercity main roads).

Graduate school began in March 1946. I was asked to visit Dean of Students Harold E. Lobdell ("Lobby"). He asked me a number of questions including why I felt I should receive one of these units in preference to someone else. I replied that my alternative was to commute from Lowell each day where we would be living with my in-laws — a long commute but that I had done a similar commute as an undergraduate from Lawrence and it was indeed possible to do it again. Shortly after that visit with Lobby, I received a letter from him notifying me that we would be assigned unit No. 72 Westgate (as the place was called), but that it would not be available for a few months as they were being set up at the time of his letter. Meanwhile I began my graduate studies as a commuter from Lowell on March 1, 1946.

First Year of Graduate Studies

I signed up for such courses as Advanced Biochemistry with Prof. Bernard S. Gould, Enzymology with Prof. Irwin W. Sizer, Electronics with Prof. Albert G. Hill and a course in Advanced Sugar Chemistry with Prof. Robert C. Hockett. I passed my first semester with a 3.75/5.0 rating — certainly not superb but at least I had not flunked out of Graduate School. It was the lowest term rating I was to receive over the next three years of graduate studies. Meanwhile, we soon moved from Lowell to Westgate. It was quite a wrench for Diana's mother who had become very attached to Errol, obviously. But they used to visit us once or twice per week — always with arms laden with goodies, foodstuffs, and a new outfit for Errol, etc.

In June, 1946, I was completely and honorable discharged from the Army of the United States and was put on a Reserve Status.

My research work connected with the S.M. degree dealt with vitamin C in orange juice and what happened to the ascorbic acid during storage. There are three forms of vitamin C — ascorbic acid, dehydroascobic acid and diketogluconic acid. The first two are biologically active; diketogluconic acid is not. This study provided the basis for my S.M. thesis and was presented at the September, 1947 Meeting of the American Chemical Society in New York City (at my own expense). It gave me a chance to attend the Free Synagogue on Rosh Hashanah and hear the great preacher, the late Rabbi Stephen S. Wise. He was superb, with a deep and expressive voice.

During the course of my S.M. program, Prof. Harris hired a Dr. Hazel Munsell to assist him in vitamin analyses. She and I just did not see eye to eye and I was forced to work as she directed or drop my assistantship. I chose the latter course, even though the lack of money meant going deeper into our savings. In the midst of all this, Prof. Harris and his group were transferred to the new Department of Food Technology headed by Prof. William L. Campbell. Prof. Proctor was the No. 2 man and Director of the Samuel Cate Prescott Laboratories of Food Technology.

I spoke with Professor Proctor in late 1946. I was then doing quite well in graduate school and Diana and I had agreed that I should go on for the doctorate. My choice was the newly burgeoning field (although not a new field) of Food Technology. All of the advances that had been made in the ancillary fields of chemistry, electronics, physics and engineering were having a major impact on the food technology of the future. (In another book which I wrote recently entitled ..."Of Microbes and Molecules," which was published in 1995 by the Food and Nutrition Press of Trumbull, CT, I have covered the history of the Department of Food Technology and term the years of which I speak and the following decade or so, "the Golden Age of Food Technology at M.I.T."). I wanted to be a part of the great advances being made by Prof. Proctor and his colleagues. I discussed my problems and my dreams with Prof. Proctor who was most generous of his time and help to me. Prof. Proctor agreed to accept me as a Doctoral candidate after I received my S.M., scheduled for June, 1947. I received a letter of acceptance for doctoral studies in Food Technology from B. Alden Thresher, dated January 27, 1947, "with no deficiencies" — a "clean" acceptance. Quite often, students who were accepted for doctoral work were accepted on the basis that they make up certain course deficiencies either at M.I.T. or elsewhere, inasmuch as M.I.T.'s program was more rigorous than that at most other universities, especially in the fields of mathematics, physics and engineering.

Birth of Judith Ann Goldblith, June 8, 1947

Meanwhile, Diana was once again pregnant. We wanted another child or two, but I had been warned by the docs on release from prisoner of war status that, in all probability, I would never again have any children, what with my past history of undernutrition and malnutrition, I and most of my fellow POW's should all be sterile. Wrong again!

Diana became pregnant and was due in early June, 1947 according to her physician, Dr. Daniel Hindman, a former Lawrencian, who had been in the Navy and had just completed his residency in obstetrics and gynecology at the Boston Lying In Hospital under Dr. Harold Rosenthal. Danny Hindman was the first cousin of Harold Hindman, who had introduced Diana and me. Diana had an uneventful pregnancy except for one bad episode which scared us. She had some type of a "bug" and Danny was afraid she might lose the baby, so he came over to the house at 72 Westgate and Diana's mother stayed with us for about a week. When Diana informed me she thought she was in labor, I called Dr. Hindman's office and we set off for the hospital. All went well except for my getting lost in Brookline enroute to the Richardson House of the Lying In Hospital. On June 8, 1947, early in the morning, Diana gave birth to a beautiful baby girl, who was named in proper form, the following week at the Temple Kehillath Israel on Harvard St., Brookline.

S.M. Degree in Food Technology

On June 13th, 1947, I received my S.M. degree at the M.I.T. Commencement exercises held at Symphony Hall in Boston. Poor Diana could not attend (that was before the day of pass-through deliveries). She was told that she would be separated from the baby until they both left the hospital in a few days thereafter if she went outside their maternity area. She just could not bear the thought of being separated from Judy. My parents and Diana's parents attended the M.I.T. Graduation and were in the first balcony just overlooking the stage. They must have arisen before dawn and my folks rushed to Lowell and Diana's father rushed them all to Symphony Hall. With doors opening at 9:30 AM, they really had to get up early to be the first in line. My S.M. degree, as my S.B. degree, was signed by Karl T. Compton who also awarded us the degrees, personal handshake and all.

No.72 Westgate

This story would not be complete without a few words on Westgate. All the units at Westgate were one story, prefabricated one- or two-bedroom bungalows — using two units adjacent to one another. They were nicely laid out, poorly constructed, and most roofs leaked in the rain, and one could use up all the pots and pans to catch the drippings from the leaks in the roof. Each unit had a small "Pullman" kitchen with a small refrigerator which made it hard to store the formula bottles. The units had no cellars but were on a concrete slab. They had a central gas heater which was really insufficient in the winter time, so without a cellar and with insufficient heat, some of us developed chilblains (nothing new to me). Each unit had a screened-in porch. But, I do not denigrate any of this, but rather praise M.I.T. for taking the initiative in making available low-cost housing for some of its veterans. Later, it had built barracks-type-construction apartments to the west of us known as "Westgate West."

Moreover, all of us were in the same boat — students with little excess cash and most with children. We saw several couples break up because of the vicissitude of living as we did with money mainly going out and but little coming in. Diana and I were lucky. We both had to adapt once before and adapted again here, especially she. We were not living in the laps of luxury but we had each other — something we missed for four years!

We had the wire-haired terrier, Butch, with us but we had to get rid of him because he was jealous of the new baby. I was told that he was no problem with Errol but we had to give him away after Judy came home to us. It was a sad day to give Butch up. He had recognized my recorded voice over the airwaves from Japan and jumped all over me in happiness when I arrived home after the war.

The Trip up Mount Washington, Summer, 1946

During the summer of 1946, Diana, Errol and her first cousin, Gail Goldwasser, and I took a trip to New Hampshire with the high point to be the cog railway ride to the summit of Mount Washington. We were nearing the summit when the passenger car, in which we were seated, and which was being pushed up the mountain by the Cog locomotive, was hit by a flatbed car, containing the garbage from the summit restaurant, that had broken loose and struck us with the momentum of the velocity of two hundred yards of a steep incline. We were lucky in that the cog held, but it was extremely windy and we had to climb the last 200 yards or so on foot up the trestle. I carried Errol, wrapped in a sheet. Gail had suffered a broken nose and Diana probably a miscarriage according to Dr. Jack Chamberlain who saw her the next day at M.I.T., because she also had a bad cut on her leg which needed attention. We were lucky because, had we been 100 yards farther down the trestle, the momentum would have been sufficient to knock us off the trestle. It was quite a struggle in the wind to walk up the trestle, but fortunately when the workers at the summit heard the accident and saw what had happened, they rushed down to help the passengers. Diana and Gail were assisted up to the top. We were all dazed as we did not know what had happened.

This was the first accident on the Cog Railway of Mt. Washington in many, many years and there had been only one previously. We were taken to the hospital in Berlin, N.H. by cars, ambulances, hearses, etc. Gail was taken in a hearse. She remained in the hospital overnight and Diana (and her) Uncle Max Greenberg drove up to get her the following day. We stayed overnight in some motel and then returned the very next morning.

So much for vacations!

Beginning of Doctoral Work and a Few Words about Prof. Bernard E. Proctor

As soon as I completed my S.M. Degree, I began to work on my Ph.D. degree program. It would be well here to interject a few words about Prof. Bernard Emerson Proctor, a remarkable individual, who became my mentor and close friend.

Professor Proctor was born on May 5, 1901 in Malden, Mass., graduated from Malden High as Class Salutatorian, M.I.T. in 1923, majoring in Industrial Biology, the forerunner of Food Technology. As a graduate student under Dr. Alan Winter Rowe, he worked as an Instructor in the Biochemistry Dept. at Boston University Medical School. He studied food technology under Prof. Prescott. He thus began an illustrious career in this now nascent field of Food Technology and built it up to be one of the best if not the best programs in the U.S. He was a superb teacher and had a great deal of industrial experience as a consultant to many large food firms in many facets of the broad field encompassing food technology.

Although I had met him before the war, while I was an undergraduate, it was only after I became a graduate student that I really began to know him, understand him and love him. During World War II, he was an Expert Consultant to the Army Quartermaster Corps, and ran a large research and development program at M.I.T. on dehydration, packaging, and he pioneered the use of ionizing radiations for the preservation of food supplies. This was the research program in which I was really interested.

He was the most unbigotted man I had met until that time. He believed that an individual needed religion. It did not matter what kind. He had a profound influence on me, my family and my subsequent career.

I could cite many stories about him and his kindnesses to many and his religious tolerances and belief in religion. Two will suffice. He was brought up in the then predominantly Jewish section of Malden, Suffolk Square, so he knew about the various holidays we, of the Jewish Faith, celebrated. One Yom Kippur, during the war, he noticed one of the younger staff members of the Jewish faith in the laboratory working. He asked him what he was doing in the lab on that day. The young man replied: "Working, Prof." Prof. Proctor said to him, "like hell you are. I don't care whether you go to a synagogue or not, that is your business, but you are not to show your face in here until Yom Kippur is over, out of respect to your co-religionists." Bernie was an active member of the Baptist Church in Melrose.

Bernie and his wife, "Pat," spent a number of seder meals at Diana's parents home in Lowell and had many meals at ours as well.

His philosophy on gratitude from students was to help if he could. One day while working on my thesis in the lab, I received a call from Diana that one of our children was ill and that we should bring her over to the doctor's office in Brookline ASAP. I stated that I would be home within five minutes and pick her and the two children up and go to the doctor's office. Prof. Proctor saw me rushing out of the lab with my jacket on and asked me if something was wrong. I briefly told him where and why I was going and his reply was. "Do you need any money"?

Another time, he had done me a great favor by bringing me to the attention of someone who employed me for some analytical work, which resulted in extra cash, which we could use as our savings were being depleted rapidly. I thanked him and stated that "I hope I can repay you some day." His reply was "some day you will be a professor and have graduate students. You can best repay me for any perceived kindnesses by doing the same for your students when the time may come." That was Bernie Proctor.

One last story about him relates to the time, near the end of my Ph.D. studies and we were fast becoming broke (fellowship, G.I. Bill, savings and all) and I told him that I might have to quit for a while and earn some money by full-time work. He said "Stick it out. If you need some money, I'll loan you some. Just remember that a Ph.D. means nothing if you have it. It only means a lot if you don't have it." How true that is, even today.

Subjects, Faculty and Research for Doctorate

The Department of Food Technology, when I began my doctorate work, was newly located in Bldg. 20, where radar had been developed during World War II. The building had been rehabilitated and provided fairly decent quarters and laboratory facilities, albeit that wooden floors do not lend themselves to chemical work nor food processing pilot plant work. We still had the use of Bldg. 35 for the Dehydration Pilot Plant. But, Bldg. 20 still stands today and is still in use.

The faculty included, in addition to Prof. Proctor and Prof. Campbell, Prof. Harris and Assistant Prof. Henry Sherman who taught courses in nutrition and did research in this field, Associate Professor Cecil G. Dunn who taught industrial microbiology, Associate Professor Ernest E. Lockhart who taught courses in food chemistry and flavor chemistry, and who pioneered the first program in flavor research at M.I.T. The engineering courses were taught by Associate Prof. August L. Hesselschwerdt, Jr., of the Dept. of Mechanical Engineering, who was an expert in refrigeration engineering.

Another important person was Ms. Elsie A. Wilson who was an assistant to Prof. Proctor and in charge of writing reports for our research sponsors. Thus many of us graduate students came into contact with her. She taught me the English language as it should be written and spoken. She had graduated from Radcliffe years earlier and had worked as an assistant to Prof. Benedict at Harvard Medical School before she came to work for Prof. Proctor during the war to take up the massive project of writing the reports to the OQMG on Army Ration research done in the Food Laboratories. She had majored in the classics at Radcliffe, and when she retired to the family homestead in South Dartmouth, Mass., she became an active member of the Great Books Club there and remained active into her nineties until she fell and broke her hip. If I have ever written anything worthwhile and in decent English, it is due to Ms. Wilson's teaching and patience.

Research Work for Ph.D. Thesis

My research work centered around the principles and use of electromagnetic radiations for the preservation of foods. This area of food research was just beginning and M.I.T. had excellent facilities in the Dept. of Electrical Engineering under Prof. John G. Trump, who made his facilities and his own assistance available to us.

In addition, Prof. Proctor had obtained an early model of the Raytheon Microwave Oven. When the Trump/Van de Graaff Electrostatic Generator was down for extensive overhaul, I worked with the Raytheon Microwave Oven and was able to study the effects of microwave energy on several vitamins and several species of bacteria, and was able to show that the effects were not due to some magic bullets as some individuals had published, but due solely to that attributable to heat. This work was published, as were the other studies done for my Ph.D. degree research, jointly with Prof. Proctor.

While we were waiting for Dr. Trump and his associates to finish the extensive repairs on the Van de Graaff Accelerator in Bldg. 28, we managed to locate a small unit which could provide sufficient soft X-Rays to destroy microorganisms in a reasonable period of time, and set out to develop dosimetric techniques and then studied the destruction of microorganisms with Soft (Kilovoltage) X-Rays.

In a few months Dr. Trump's Van de Graaff was rebuilt and we began to study the effects of fast (Supervoltage) electrons on microorganisms, vitamins, amino acids, in pure solution and in situ. Those were exciting days and everything (almost) was publishable. We published a series of papers in a new journal, Nucleonics and papers in the Journal of Biological Chemistry, Bacteriology, Journal of Applied Microbiology, Food Technology, Food Research, etc.

In 1948, I sat for and passed my written and oral examinations for the doctorate. That was an ordeal! I also completed my minor requirements in biochemistry and passed my language examinations in French and German. In January, 1949, I successfully defended my doctorate thesis and was awarded the Ph.D. signed by James Rhyne Killian as "President Designate" — one of the few degrees which he signed with that appellation.

In June 1949, I attended the graduation at the Rockwell Cage to receive my doctorate hood. This time, Diana was there as were both our sets of parents. Bernie Proctor took some colored photos of me in my regalia. It had been quite a struggle for Diana and Errol. Studying at home in quiet was necessary and Diana made sure that was the case and Errol put up with it. In fact, one morning he got up early at Westgate (so what else is new) and we heard him exclaim: "Daddy must have finished his thesis!" I had cleaned off my desk in the living room before I retired the evening previous.

Ms. Wilson helped me in polishing of the writing of my thesis. The papers that resulted therefrom appear amateurish and simple today, but at the time were considered to be the seminal works in the field (and probably still are), although I have not kept up with that field for some years — even before I retired. I consider myself as a generalist, and after, the pioneering work had been done on "cold sterilization." When the Van de Graaff Accelerator in Bldg. 28 had been rebuilt, I went on to other interesting problems.

I helped start the sterilization of medical supplies and surgical sutures. I also sterilized the aorta from dogs, which, until the advent of good plastic, cross-linked polyethylene tubing, was used for repair of aortic aneurysms. This surgery was pioneered at Children's Hospital by Dr. Robert Gross. These were exciting days, wherein I was participatory in and a witness to the birth of a whole new industry. I got to know Prof. Robert J. Van de Graaff, Denis Robinson, the founding President of the High Voltage Engineering Corp., as well as Prof. Trump (who had been Van de

Graaff's student and reduced the size of the Van de Graaff accelerators by introducing pressurized gas, instead of air, for insulation). The original financing of the H.V.E.C. had been through the American Research and Development Corp., a pioneer venture capital firm established by Gen. George Doriot, then of the Harvard Business School and Dr. Karl T. Compton, then Pres. of M.I.T. H.V.E.C. was among American Research and Development Corporation's first success story, along with Digital Equipment Corp. and Ionics.

The High Voltage Engineering Corp. had been formed by Van de Graaff, Trump and Robinson to produce Supervoltage, deep therapy, machines for cancer treatment. This was a field in which Trump was one of the pioneers. Later the company built tremendous Van de Graaffs known as the Emperor Van de Graaff and later, even larger ones, Emperor Tandem Van de Graaffs, for studies on the structure of matter. When the research funding for physics research diminished greatly from the U.S. government, Trump had already set up new ideas of cross-linking plastics by bombardment with fast electrons and had built up an H.V.E.C. subsidiary known as Electronized Chemicals Corp., adjacent to H.V.E.C. on Rte 128, in Burlington. The company had its ups and downs but mainly ups. I was honored in about 1979 when Drs. Trump and Robinson called on me to inform me that they had reached their mandatory retirement age from the board of H.V.E.C. and wanted to ask me, on behalf of the Board, to assume their place on the Board of H.V.E.C. When the company was acquired, in the mid-1980's, it was doing extremely well due in large measure to the then C.E.O., Dr. Pascal Leveque, whom Dr. Trump had brought in originally to run Electrolyzed Chemicals. Dr. Leveque is a superb executive and a knowledgeable electrical engineer.

Almost as soon as I had completed my doctorate studies, Diana and I received notice that, given a reasonable amount of time, we would have to leave 72 Westgate. There was a big waiting list and I was not going to be a student at M.I.T. much longer. We could not afford to buy a house, but we were able to find a suitable house on Laurel Terrace not too far from Union Square, Somerville, to which we moved in the spring of 1949. We had wonderful neighbors there, Hazel and Jim Busby with whom we soon became friendly. Their middle daughter, Rita Busby (now Lenox) baby-sat for us periodically. The commute to M.I.T. was only about 10 to 15 minutes. It was a good move for us at that time.

In 1948-49 I began to consult for several companies, including Arthur D. Little, Inc., for whom I consulted exclusively during the decade of the 1950's. We were able to buy a new car — a Studebaker sedan, I recall.

In my last year of doctorate studies, Diana's father thought it would be a good idea if I joined the Masonic fraternity and arranged for me to join a military lodge in Boston, the Major General Henry Knox Lodge and he paid my initiation fee. Dad thought that I should join early in life before I made too many enemies and could be blackballed! I joined the Major General Henry Knox Lodge and shortly thereafter learned that there was a Masonic lodge at M.I.T., the Richard C. Maclaurin Lodge. Prof. Proctor asked me why I had not applied there. I informed him that I did not know, at the time I applied to the Major General Henry Knox Lodge, that a similar lodge existed at the Institute and accepted only those with a connection with the Institute. I learned from Prof. Proctor that he had been Master of the Lodge as had Dr. Vannevar Bush who had been the second master. He could not attend my raising to the Master Mason Degree because of travel, but gave me his own Masonic ring which I have to this day. I joined the M.I.T. lodge about 48 years ago, went to their meetings and later, in 1957, I became the Worshipful Master, thus following an illustrious group of M.I.T. stalwarts.

Friendly Professional Acquaintances, New Friends, a New Job

During my pre-doctoral studies, I got to know Dean Prescott very well, he was then in his 70's and very active in writing and in the Institute of Food Technologists (now with a membership of over 25,000) and in The Refrigeration Research Foundation. He used to autograph his books to me as either "Your Fellow Food Technologist" or "The Elder Sam." He was a delightful gentleman. We became even closer when Professor Proctor passed away in September, 1959. Through Dean Prescott, I met Mr. James Dole, one of the co-inventors of the Dole Aseptic Canning System.

I also got to know very well, all of the major technical leaders of the American Can Co., the Continental Can Corp., Wilson and Co., Armour and Co., General Mills, Pillsbury and others. These people were always coming to M.I.T. to learn more about what we were doing and were giving financial support to our work. Dr. Charles N. Frey, the Scientific Director of Fleischman Brands was very friendly and later lectured in our Department — which Prof. Proctor arranged and I continued until Dr. Frey chose to retire, many years later.

Through Prof. Proctor, I was introduced to Prof. Samuel Lepkovsky, the eminent nutritionist, who single-handedly won the race to crystallize vitamin B_6 and later discovered and identified Xanthurenic Acid in the urine of vitamin B_6-deficient rats. We became very good friends until he passed away. Our paths crossed in Edinburgh, London, San Francisco and even Tel Aviv as the years went on. Sam Lep came to dinner at our house at 72 Westgate when I was a graduate student, and I recall that Diana served him a Kosher hot corned-beef meal which he relished very much. We would often get together at the Proctor's camp on the Contoocook River in N.H. or at their apartment in Cambridge. He was a great guy and a connoisseur of fine wines — and good people. Through Lep, we met D. Wayne Woolley, the Ostrers, Sandy and Louise Feldman, Irving Sharon, etc. More about these later.

First Post Doctorate Position

Prior to the completion of my doctorate studies, I was asked by Prof. Proctor and Prof. Campbell to stay on at M.I.T. as a

Research Associate. This suited Diana and me very well even if the pay was not what I could earn in industry, as we did not have to move, would be near family, and the potential was there, if I did well, for a bright future at the Institute. Moreover, I could continue my research and do some teaching — both of which I enjoyed. So the decision was made that was to guide my future career — remain at M.I.T.

CHAPTER 13

RESEARCH ASSOCIATE AND ASSISTANT PROFESSOR (1949 - 1955)

Personal

Diana and I decided that we should develop roots in a community within commuting distance of Cambridge, with a good public school system and good public transportation. We selected Melrose and looked for a house to rent. We planned to purchase something later. We rented an old house, in 1951, on Upham Street and Errol attended the Winthrop School, which had some very good teachers and an excellent principal, Mrs. Hawes.

We joined the Melrose Temple which rented space both for High Holyday Services and for the Sunday School at the American Legion Bungalow. Later, they purchased their own building on West Foster St. Through the Temple we met Dr. and Mrs. Miah H. (Gerry and Esther) Rovner, Sidney and Florence Lewis (Sid and Florence) and the Samuel (George and Bea) Whites with whom we became very friendly and have remained so, although Diana, Florence, Gerry, and George have since passed on. Our medical needs were first taken care of by Dr. Samuel Norman of Malden, father of Drs. Leona Zarsky and Paul Norman, both schoolmates at M.I.T. Unfortunately Dr. Norman, a wonderful man and a superb physician, passed away relatively early on. Although he specialized in obs-gyn, he did not do so exclusively because when he had to make the choice in the 1930's, the birthrate in the U.S. was decreasing and he had two children to send through college and medical school. For our dental needs we continued to go to Brookline to Dr. Bertram Fogel on Commonwealth Ave. until Sid Lewis introduced us to Dr. Jacob (Jack) Glazin in Malden and later his son, Dr. Allen. Jack or "J.J." was a real extrovert with many outside interests, including the masonic fraternity and Temple Tifereth Israel in Malden (to which we still belong, as do our children). We wanted a real comprehensive Hebrew School program for our children which Errol did not get in Melrose and the Temple in Malden provided.

Birth of Jonathan Mark Goldblith

Our third and last child was a son whom we named after my maternal grandfather. He was born at the Malden Hospital on July 7, 1953. Dr. Samuel Norman delivered him, an 8 pound 14 ounce strapping boy with reddish curly hair. He looked ready to enter kindergarten! Jonny was the baby of the family including his cousins. As the youngest, he had to ask the four questions at the Passover Seders until our granddaughters, Sarah and later Rachel, took over. Rachel, at age 21 still does it at the time of this writing, Spring, 1996.

Jonny was always a bright, sunny boy and always getting into mischief as a boy. He continues to give us a great deal of pleasure, but today it is through his professional accomplishments and his voice. But, he disliked school — due to a problem with one teacher.

After Dr. Norman passed away in 1953, shortly after Jonathan's birth, we turned to Dr. Rovner for our medical needs.

In 1953, we purchased our first house on 76 Norman Road in Melrose with the aid of the G.I. Bill, 4% Mortgage loan plus an additional loan of $700. from Florence and Sid Lewis. The house had but three bedrooms and we had one bedroom and a large closet modified to make a separate bedroom for Jonathan, so that each youngster had his/her own room.

Bar Mitzvah of Errol David Goldblith

We had a special Hebrew tutor for Errol for his Bar Mitzvah Training. On a snowy Saturday in February 1955, we celebrated Errol's Bar Mitzvah at the American Legion Bungalow cum Synagogue. Our parents stayed at our house along with the cantor we had imported for the services, Cantor Morris Lewin. We had a kiddush after the services and then a proper dinner, the next night at the M.I.T. Faculty Club, which permitted us to bring in a Kosher Caterer (the first time in its young history). Errol did his "Maftir" beautifully but it was a trying event for us all. Poor Diana had to cook for the Friday night preceding for the Rabbi we had imported, who stayed at a neighbor's house across the street, while we stayed at the Rovners. But it was a simcha to be remembered with joy and pride, as was the Faculty Club affair.

FIG 13.1. ERROL DAVID GOLDBLITH'S CELEBRATORY BAR MITZVAH PARTY, FEBRUARY, 1955
Steven Greenberg, Barbara Greenberg, first cousins, and Errol on the right. At M.I.T. Faculty Club.

FIG. 13.2. KAY GREENBERG (DIANA'S AUNT) DANCING WITH DIANA'S BROTHER, IRVING GREENBERG, AT ERROL'S BAR MITZVAH PARTY

The next year, however, we joined Temple Tifereth Israel in Malden, on the advice of "J.J." Glazin, and Judy and Jonny, as well as Diana and I benefited in a number of ways.

Golden Wedding Anniversary of Abraham and Fannie Goldblith

On Sunday, March 19, 1954, Diana and I had a catered dinner and reception for my parents at the M.I.T. Faculty Club in honor of their fiftieth wedding anniversary. We had all of their friends, Diana's parents, uncles and aunts, our close friends and colleagues at M.I.T., Rabbi and Mrs. Joseph Warren of Lowell, who married Diana and me, and Rebbe Sol Golubzik, my revered Hebrew School teacher, who came with his daughter. Naturally we had a kosher caterer. Bernie Proctor and Pat were there and livened up the place quite a bit. My father wore a vest my mother had made for him for their wedding. It was a festive occasion, needless to say.

FIG. 13.3. PARTY AT M.I.T. FACULTY CLUB IN HONOR OF MY PARENTS' FIFTIETH WEDDING ANNIVERSARY, MARCH 19, 1954 — THE HAPPY COUPLE!

FIG. 13.4. AT FIFTIETH PARTY, FRONT, MY MOTHER AND FATHER; REAR ROW, DIANA'S MOTHER, FATHER, DIANA AND I

FIG. 13.5. RESPONDING TO THE TOAST, MA AND PA

FIG. 13.6. DIANA WITH RABBI AND MRS. JOSEPH WARREN AT FIFTIETH PARTY

Professional Activities

My research work continued on from February 1949 as a Research Associate. In 1949 or 1950, I was given an opportunity to do some teaching. I handled the laboratory course in bacteriology for Prof. Cecil G. Dunn. I was also given the opportunity by Prof. Proctor to develop a new graduate course in Food and Drug Applications of Ionizing Radiations.

Professor Proctor and I wrote, in 1951, a major review article entitled: "Radiation Preservation of Foods and Drugs" for the then new series of books, *Advances in Food Research, Vol. 3.*

My research continued to go well and I began to get some help from new graduate students and some research assistants. There were some individuals — students and faculty members — about whom I want to say a few words further on.

FIG. 13.7. MY MOTHER DANCING A "FREILACH" WITH JENNIE ISSENBERG AT THE FIFTIETH

Among the notable consulting work I did during the period herein covered was that with Arthur D. Little. In my work with them, we developed the most powerful three million electron volt Van De Graaff Accelerator then built. It had two tubes, scanning 12 Kilowatts of 3 MEV electrons used for the sterilization of vegetable material for a client of theirs. It was set up in a pilot plant in North Cambridge on Huron Avenue. The scanning of intense beams of electrons was developed by this group by analogy with a TV tube which was just becoming very important commercially in the U.S. then, in the early 1950's. This consulting work with ADL continued until circa 1961. Through it, I met General James Gavin who became President of ADL, and we became good friends until his death.

Two Honors and Awards in 1953

On February 19, 1953, I had an unexpected honor. I was selected by the Boston Junior Chamber of Commerce as one of the Ten Outstanding Young Men of Boston for 1952. I was very surprised inasmuch as the awards of the previous year included John F. Kennedy, later to be elected President of the U.S.A., and Edward Brook, later to be elected U.S. Senator from Massachusetts. Also selected in 1952 were Paul Cusick, assistant Director of Industrial Corporation at M.I.T., later to be Vice President of Finance, and Dr. John Merrill of the Peter Bent Brigham Hospital and a pioneer in the development of kidney dialysis. I learned later that I had been nominated by Prof. Proctor.

In 1953 also, about a month before Jonathan was born, the Annual Meeting of the Institute of Food Technologists took place in Boston. I delivered a paper on the radiation preservation of foods. There was to be a prize offered by the Monsanto Chemical Co. for the best paper of the meeting. For reasons I cannot fathom to this day, our paper (Proctor's and mine) which I delivered, was selected as the best and I received a check for $300.00 plus a plaque presented to me at the Fall Meeting of the Northeast Section of the I.F.T. by Mr. James Donovan, M.I.T. 1928, Chairman of the Northeast Section of the IFT. Diana was there, but she could not attend any of the I.F.T. meetings in June.

In 1953, Prof. Emil M. Mrak, then Head of the Department of Food Technology at the University of California, and later scheduled to become the Chancellor of the University Campus at Davis, spent a sabbatical semester at M.I.T. as a Visiting Professor. He talked me into coming to California the following year when the I.F.T. was scheduled to hold its annual meeting there in June.

Diana, Errol and I drove out in an almost new un-airconditioned Buick. Diana's mother, Bless her Heart, volunteered to stay with Jonathan and Judy. We tried to get Judy to come along as we were sure that she would enjoy the national parks, etc. Judy would have no part in it. She preferred to stay home and be with

Grandma, Jonny and her friends. She hated long rides and hasn't changed in that regard to this day.

FIG. 13.8. MY MOTHER (L) AND DIANA'S MOTHER (R) AT OUR HOME ON NORMAN ROAD, MELROSE (CIRCA 1954-5)

FIG. 13.9. MY PARENTS AT NORMAN ROAD, MID 1950'S

First Family Trip by Automobile to California, 1954

Diana, Errol and I had a great trip out to California, visiting some tourist spots, national parks, friends, etc. We first went to San Francisco having stopped at the Bad Lands, Yellowstone (snow), the Grand Tetons, Reno, etc. At San Francisco, Prof. Sam Lepkovsky took over part of our schedule. We visited the Mraks in Davis with him and there, Emil Mrak took us to Louis Martini's Winery and we were shown around by Louis P. Martini, the son of the founder, who had been a student of Emil's years earlier. His father, the founder, Louis M. Martini, entertained us with fine wines in his cave and sampling room. (I still get a discount from the winery whenever I go there.) It was difficult to stay awake in the June heat and to drive further, but drive on we did to the Inglenook Winery, and finally to Berkeley where we had a motel room. Lep took us to some of the finest restaurants in San Francisco. Ernies was a notable one, as was "Sam's Grill" which specialized in seafood dishes, and where Lep was well-known as "The Professor" and "Two $B_{6's}$ make a B_{12}." I don't know how many times I went there afterwards with him, and later by myself or took friends there. It is still one of San Francisco's great culinary establishments. Their "Rex Sole a la Sam" was among the unusual dishes and their sourdough bread was outstanding.

In San Francisco, I had the pleasure of introducing Diana and Errol to the Sygmund Habers who had been so nice to me on the way overseas. In Berkeley, I also was introduced for the first time to Irving Sharon, a dental surgeon who worked in his spare time with Lep, and to Dr. Sanford Feldman, a senior surgeon at the Mount Zion Hospital in San Francisco, who also worked with Lep and who later was to become the President of the San Francisco Medical Society. Both of these men became close friends of ours and were real "menchen." Sadly Irv Sharon has since passed on.

Then we drove to Los Angeles to the I.F.T. Meeting and to the usual touristic spots. Then we began the long trek back to Massachusetts, beginning with the desert (where Diana almost passed out from the heat), along to the Grand Canyon, Zion

National Park and across the U.S. on Rte 66. I recall that we stumbled into a big Indian Pow Pow in Williams, Arizona, and that I almost burnt my mouth on Mexican food in Albuquerque.

More About The Radiation Preservation of Foods Program

In the period between 1949 and 1953, I became friendly with Dr. Peter Demos, who designed and built the first M.I.T. Linear Electron Accelerator, and this afforded me an opportunity to extend my work on the radiation preservation of foods to the 15 MEV energy level.

In 1953, also, the QMC began its famous program on Radiation Sterilization of foods following the feasibility study by Dr. Ralph G.H. Siu and a new vista opened up for me. Prof. Proctor was the Chairman of the National Research Council Committee on Radiation Sterilization of Foods which had oversight on the Army QMC/Surgeon General R & D program on developing field rations for the Armed Forces using Radiation as the means of Preservation. I was asked to be the Technical Aide to the Committee and later, well after Prof. Proctor's death in 1959, I became the chairman of the committee. This Army program was initially an external one and broadened the numbers of institutions working in the field from just Electronized Chemical Co., Swift and Co. and us, to many, many more. Work began under the Office of the Surgeon General of the U.S. Army on the wholesomeness of irradiated foods.

This QMC program, plus the program under the aegis of the Atomic Energy Commission, broadened our own sources of funds and research projects at M.I.T. We employed Dr. Sanford A. Miller, a newly-minted Ph.D. from Rutgers to oversee our work on the wholesomeness aspects of the program. Sandy had come highly recommended by Prof. James Allison of Rutgers, the renowned protein nutritionist, whose first graduate student, years earlier, had been Dr. Irwin Sizer, my first Instructor in Biology at M.I.T., later to become Head of the Department of Biology and Dean of the Graduate School at M.I.T. It's a small world!

The irradiation of foods program broadened and became international with laboratories from Japan, Germany and the United Kingdom becoming involved. Thus, I made more friends and gained graduate students and post-docs from Japan and England.

Promotion to Assistant Professor

In March 1952, Prof. Proctor succeeded Prof. William L. Campbell as Acting Head of the Department and about nine months later was appointed department head. In 1953, both John T.R. Nickerson and I were appointed Assistant Professors.

Staff with whom I Became Friendly in the Early Days

Prof. John T.R. Nickerson was an alumnus of the Institute and had done his undergraduate and graduate work with Professor Proctor. He was a most interesting individual — a hard worker, and Nova Scotian by birth, who had gone into the food industry after he had obtained his doctorate and had several areas of particular interest. These were food bacteriology, fisheries technology, and freezing preservation of foods. He had been in the packaged frozen food industry from its earliest days and knew the people, products, and processes. He was a great guy also.

Nick came to work at M.I.T. on Feb. 1, 1948 as a research associate. My doctorate orals were on a Saturday afternoon and Nick spent the morning with me, keeping me calm, took me for a snack at Walker Memorial before the orals. He did quite a bit of research, as well as undergraduate and graduate teaching at M.I.T., and was an important part of the special summer program course instruction for industry. He and I were both promoted to assistant and associate professorship at the same time, and remained good friends until his retirement and afterwards as well.

Graduate Students as Friends

I have had some interesting graduate students, a number of whom became close personal, as well as professional, friends. It is dangerous to try to mention them for fear of leaving out some but I will do so anyway. The first one that comes to mind is the late Dr. Yaichi Ayukawa of Tokyo, Japan. We have become personal friends with him and his lovely wife, Masako and their son Junta, who was born in Boston. Yaichi became an important Member of the M.I.T. Corporation in the late 1970's, and we received the Bronze Beaver Award of the M.I.T. Alumni Association at the same time on the morning of the Inauguration of Dr. Paul Gray as President of M.I.T. in 1980. Further on, I will relate some of Yaichi's activities regarding fundraising in Japan and the opening of the Industrial Liaison Program office in Japan.

Dr. Richard I. Mateles, who later became a member of the M.I.T. Faculty, as well as the Faculty of the Hebrew University in Jerusalem, and his wife, Rosalind, became (and still are) close friends. The late Dr. Harold M. Rich was a former thesis student and we enjoyed very much the fishing trip we went on with his wife, Cornelia (Sunny). Hal Rich worked for the Kitchens of Sarah Lee and through him I also met Charles Lubin, the founder of this famous company that produced such quality frozen food products.

During the 1950's and 1960's we had a number of Quartermaster Corps officers. We became particularly friendly with two of these officer/students. The first was the late Major Reuben Pomerantz and his wife, Harriet. The second and whom I still see and hear from regularly is Colonel (then a Captain) Rufus Lester and his wife, Martha. Rufus has been just great to me and I am proud of his accomplishments. After he retired from the army as a full colonel, where he was in the position of Commanding Officer of the Natick Laboratories of the Q.M.C., he became Vice President and later was promoted to Group Vice President of The Stop and Shop Companies. Rufus retired from Stop and Shop a few years ago.

I would like to mention four other former graduate students with whom we became particularly friendly (as well as with their wives), and one former staff and faculty member. These are Steven Tannenbaum and his wife, Carol. We attended their wedding in

Cedarhurst, L.I. on Labor Day, 1959. Our son, Jonathan, was the Cantor at their daughter Lisa's Bat Mitzvah, and later was the Cantor at Lisa and John Hanselman's wedding in 1985. Steve is a first-rate food technologist and a brilliant chemist who now is a faculty member of the M.I.T. Chemistry Department and will become head of the Toxicology Division of Whitaker College on July 1, 1996. A second, whom I have not seen too much recently because of distance, is Graham Lusk and his wife, Sally. Sally formerly worked for Prof. Robert Harris when she met Graham in the Department. Diana and I were at their wedding also.

Marcus Karel and I go back many years and worked together beginning in the 1950's. Marc is an unusual person who survived the Holocaust and studied at the Technical University of Munich for a while and then came to the U.S. with his family. He began work at M.I.T. as a pair of technically trained hands, went to B.U. for a bachelor's degree and then worked as a research associate in the Department, obtained his S.M. and Ph.D. degrees, and went up the ladder in the department through the ranks of full professor and succeeded me, in 1974, as Associate Department Head. He and his wife, Carol (Cal), have been friendly since our early days together.

Stanley and Shirley Charm became close friends almost from the start and we have remained that way. I hear from them frequently. Stanley's uncle was Bill Gorfinkle, a friend and close associate of Bernie Proctor's, and Stanley's father was an M.I.T. alum. Stanley was always inventing things and many of these have come to fruition and of value. He and Shirley are now in a business as a result of one of his inventions — that of detection of minute quantities of antibiotics and sulfa drugs in dairy and meat products. He has developed and built a rapid HTST sterilizer, which is very accurate and of value to the pharmaceutical industry. But mostly, they are to me dear and close friends.

Louis Ronsivalli came to the department after he received his degree in Biology from Boston University. He was (and still is) very talented with his hands as a carpenter (he built their first home himself), and was very helpful to us over the years, as well as being good friends.

Lastly, in the category of students and staff, I want to mention Judith Cohen Miller and her husband, Sanford, now Dean of the Graduate School of Medical Sciences at the University of Texas at San Antonio. Judy never forgets a birthday of mine or Diana's, wedding anniversaries, etc. She was a research assistant of Nick's when Sandy came to M.I.T. in about 1957. Diana and I helped to make the "shidduch" and attended their wedding at the M.I.T. Chapel. They have been dear and wonderful friends.

Ray and Geula Pariser are in the same category of good friends that arose out of Department of Food Technology affiliations. Each of them is unusual in his/her accomplishments. But to me, the most important is their friendship which I value very much.

Suffice it to say that one of the most wonderful aspects of being a teacher is to make friends with some of the former students and to see one's students become more able and more notable than oneself. Paraphrasing the Good Book, "A wise student maketh a proud father." Additionally, having administrative responsibilities brings the added bonus of meeting new friends, such as the Parisers and Millers, as well as many other fellow administrators.

Summer Camp at Northwood Lake

In the Spring of 1956, we looked around to possibly rent or hopefully purchase a summer camp within a reasonable distance of Melrose. Jonny was then three and it would be good to have a vacation place. With the aid of the Proctors we located Northwood Lake, N.H. and found thereon a small camp, which was available for rental for that July and August. Diana took the children up and stayed with them and I was up there for five nights per week. The commute that way was not too bad. This gave us a springboard and an opportunity to look around.

Late that summer we saw a for sale sign on a camp on the western end of the lake, opposite Route 9-202 — the Concord

Portsmouth Road, owned by a Mrs. Slater of Suncook. We purchased the camp that fall and the next year enclosed the open screened-in porch with second-hand storm windows which our friend and colleague Louis Ronsivalli obtained for us in Lawrence for fifty cents each. Several of the staff come up for a Saturday and Sunday cookout and porch enclosure; including the Ronsivallis, Licciardellos, Millers, and Gauthiers.

With the porch enclosed, I then hired a local jack of all trades, Mr. Bill Young, to raise the roof and to begin the making of a proper second floor. He put up the studs for making four small bedrooms from an open attic. I panelled the rooms upstairs with knotty pine in the next few years after I had "knotty pined" the downstairs, and Louis Ronsivalli and I improved the kitchen. In about 1984, we had a second floor bathroom installed which cost twice the initial outlay for the camp. It was all worthwhile. Now, the children and grandchildren come up and we have had a whale of a good time. Our friends the Issenbergs bought a camp not too far from us in the 1960's, and Ruth and the boys spent the entire summer there for some years.

We purchased some boats of varying descriptions from a pedal boat to a big floating porch, which we now have. It was a good investment for young people to breed togetherness. Now, a bit unsteady on my pins, I seldom get up there but Jonny comes up when he can, and Judy and Rich come weekends as often as their schedule permits.

Diana took the two granddaughters up there so that Judy could study for the bar examinations in the summer of 1980 and she came up for a day or two on the weekends. The New Hampshire bit of property has been good for us.

Two N.H. families have become close friends. Paul and Lorraine O'Neal have opened and closed the camp each year and, in addition, Paul took care of the plumbing and electricity needs. The other family is the MacCallums, from whom we have purchased our boats, and who service them and still are good enough to put in and take out the dock each year. Sandra and

Donald are just great people and both the O'Neals and MacCallums have become like family. Without them, we could not have maintained and kept the camp all these years.

CHAPTER 14

ASSOCIATE PROFESSOR (1955 -1959)

My responsibilities in both teaching and research increased. We received two new irradiation sources in the Department, a kilocurie cobalt-60 source for which we had designed and built through the Physical Plant Department, a small building located where the Landau Building is now situated. We also received a large, self-contained Cesium 137 source from the A.E.C. which was housed in the Department. It was a large kilocurie source. Thus, we had both the man-made isotopic and the naturally-occurring sources available for our irradiation program. In addition, in the mid-1950's, the General Electric X-Ray Company loaned us a One Megavolt Resonant Transformer. We were well-equipped in this field in terms of irradiation facilities.

In 1952, Prof. Proctor was stricken with a severe coronary from which he, fortunately, recovered. But he was advised to slow down a bit, lose weight, etc. This, of course, gave me more responsibilities, especially in research and administration. Professor Nickerson assisted him in his teaching responsibilities.

During this period, our undergraduate students were not increasing in numbers at all, but our graduate students census was increasing rapidly, as research funds grew and concomitantly, funding for graduate student support grew.

Only ten percent of the budget of the department came from the Institute. The remainder came from "soft" money — Army and A.E.C. contracts/grants, and in later years, N.I.H. grants and training grants funding. This was wearying on Prof. Proctor, especially inasmuch as he was well aware that other research oriented departments were getting 25 percent of their budgets from Institute General Funds.

In 1955, Prof. Proctor prepared to attend the Atoms for Peace Conference in Geneva where he was to present one of the key papers on food and drug preservation by irradiation. Prior to his departure, at his request, President Killian named me as Executive Officer of the Department. Prof. Proctor wanted someone as a "second-in-command." who would also be in charge in his absence. I was flattered, of course, but I felt that some of the more senior faculty probably were not happy with my selection. This was, of course, understandable, but I had not gone looking for the added responsibilities. I was heartened however by a personal note I received from President Killian that he was delighted that I was going to help Prof. Proctor shoulder the administrative responsibilities of the Department. The Executive Officership began a period, for me, of over thirty years of continuous administrative responsibility at M.I.T.

In 1957 or 1958, I've forgotten exactly which year, while Diana and I were visiting the Proctors at their Contoocock summer home, Prof. Proctor suffered a second heart attack. I drove him, while he was lying down in the back of our 1957 Buick station wagon, to the Concord, N.H. Hospital. He soon was well-enough to complete his rehabilitation at the M.I.T. Infirmary, where he was well looked after by Dr. James Howard Means as he was following his original heart attack. He did not slow down, but for a week or two, as I recall, and still kept very active, primarily traveling around seeking funding for the department, as well as supervising graduate thesis research in packaging, freezing of foods, irradiation, and other areas. He was a generalist and an innovator.

As I had mentioned earlier, I had joined Richard C. Maclaurin Lodge AF&AM and in late 1957 was elected Worshipful Master for the year 1958. We had an open installation in December, 1957 and Prof. Proctor and some other colleagues were part of the installing suite. It had been a very busy year, inasmuch as I had just served as Chairman of the Northeast Section of the Institute of Food Technologists for the academic year 1957-1958. My dear wife had to put up with more nights out but seldom mentioned it. All of the family were at my installation as Master of the Lodge except Jonathan, whom we felt would find it difficult not to fidget and/or to stay awake. It was an evening long to remember.

FIG. 14.1. "TABLE LODGE" HELD AT RICHARD C. MACLAURIN LODGE, AF & AM , IN HONOR OF SENIOR PAST MASTER, DR. VANNEVAR BUSH
(From l to r) Dr. Samuel Goldblith, Presiding Master, Dr. Bush, W.A. Hokanson, PM & Sec., RCML, Whitfield Johnson, Past Grand Master of Mass. Grand Lodge of Masons (April 28, 1958)

In mid-1958, I was put up for promotion to Professor by Prof. Proctor but did not receive it, I was told, because of the large number of candidates up for promotion. I had not expected to be nominated much less to receive Academic Tenure, which was awarded me effective July 1, 1959.

In 1959, we prepared and organized an International Conference on the Radiation Preservation of Foods and Medical Supplies by Ionizing Radiations held July 27-30, 1959. This conference was under the joint sponsorship of the U.S. Atomic

Energy Commission and the International Atomic Energy Agency. Through this international meeting and playing a role in its organization and execution, I was able to meet a number of new professional colleagues from around the world. The proceedings of the conference, 289 pp, which I edited, were published in November of 1959 about six weeks following the death of Prof. Proctor.

Promotion to Professor, July 1, 1959

I was again put up for promotion to Professor a year later, in March, 1959 and this time received the promotion effective July 1, 1959. Professor Proctor had written a letter to my parents, which I still have, congratulating them. He had said nothing about keeping it quiet until M.I.T. made the official announcement. They, of course, proudly informed some of their friends, who saw to it that they were interviewed by the Lawrence Eagle Tribune and a story about the Bataan Death March Survivor promoted to a Full Professorship appeared in the Lawrence daily newspaper a month or two before the M.I.T. official announcement. In the event, no harm was done. It was most thoughtful of Prof. Proctor to write my parents that personal letter. I think my promotion to a full professorship gave my father more joy than almost anything else that happened to me. It was his idea, when I was a little boy, that I should study at M.I.T. and neither of us could ever imagine that I would someday be fortunate enough to be appointed a full professor at a place we both loved.

CHAPTER 15

PROFESSOR OF FOOD SCIENCE AND TECHNOLOGY (1959 - 1974)

My duties as a faculty member and responsibilities as Executive Officer of the Department remained the same. There just seemed to be more to do. Our work on irradiation was proceeding well.

During 1958, Professor Proctor had been approached by Paul M. Jacobs, Executive Vice President of the Gorton Corporation of Gloucester, one of America's leading seafood processing companies, to join their Board of Directors. Prof. Proctor informed Paul that he appreciated the opportunity and honor, but that "his plate was full" in terms of outside activities. Moreover, much as he would like to do so, his health would not permit this added responsibility. He recommended me for board membership.

A few weeks later, I was invited to meet the Board of Directors of Gortons, and in 1959 was elected to membership to the Board. It was a most worthwhile experience for me to broaden myself in terms of knowledge of the food business, as well as in terms of meeting individuals. I became very close with Paul Jacobs and Bob Kinney, the President. I remained a member of the board of the Gorton Corporation until it was sold to the General Mills Corp. in 1968. All of this did well for M.I.T. as well, in terms of fellowships and later on, when General Mills bought out the Gorton Corp. A decade or so later, Paul and I, at his suggestion, set up a consulting partnership which lasted until his untimely death in 1980 due to cancer. He was a great guy and I was very touched when he asked me to deliver the eulogy at his funeral.

In 1960 or 1961, I was approached by Mr. Arnold T. Hampson to become a member of the Board of Directors of Merchants Cold Storage Warehouse and its subsidiary, Maritime Terminal of New Bedford, to take Prof. Proctor's place on the

Board. I gratefully accepted and remained on the Board until about 1985.

Death of Professor Bernard E. Proctor

On September 24, 1959, Prof. Proctor and I went out in his car for a sandwich lunch (delivered to and eaten in the car) at a relatively new drive-in concept known as "Richards Drive-in Cafeteria" on Memorial Drive, Cambridge. After the sandwich, French-fries lunch, we both returned to our respective offices. Shortly thereafter, I heard a commotion from the department headquarters office. Prof. Proctor had been sorting out his seminar speaker card file trying to decide whom to invite to speak when he just dropped dead.

We called the Medical Department, tried resuscitation, but all to no avail. A giant personality had left us. I called Diana to inform her and ask her to hunt up Mrs. Proctor who was, I recall, at a beauty parlor in Malden getting her hair done. Diana, with Errol accompanying her, located Mrs. Proctor and informed her of the tragic news and brought her home. By then Prof. Proctor's body was at the Infirmary and they were awaiting the Medical Examiner.

We were all shocked, of course, at Prof. Proctor's untimely passing. We knew that there would probably be great changes in the department.

Interregnum (October 1959 to September 1961)

Shortly after Prof. Proctor's funeral, I was asked by Dr. George R. Harrison, Dean of the School of Science, to serve as Acting Head of the Department until a new department head could be selected. My terms of reference were status quo — no new appointments. I have related all of this in the book published in 1995: *Of Microbes and Molecules*, loc cit and will not go through it again here. In the event, after much discussion, the dean permitted me to request the appointment of Sandy Miller and Emily Wick as assistant professors, which was approved by President Stratton without any further problems.

In early 1960, Dr. Nevin S. Scrimshaw was designated as department head of a newly expanded department, both in terms of scope and of faculty numbers. Scrimshaw would not arrive until the fall of 1961. I agreed to serve as acting department head until he arrived on the scene in August 1961. I kept the department together to the best of my ability and with the help of new and younger staff and faculty.

We began some research in 1959 on freeze dehydration of foods. A graduate student, Graham Lusk, wrote his thesis on "Astacene pigments of freeze-dried salmon during storage." Marc Karel and I jointly supervised his thesis, as I recall.

My own research, done in conjunction with colleagues, such as Marc Karel, Steve Tannenbaum and Dan Wang, had expanded more into freeze dehydration of foods, as well as on the effects of microwaves on foods and their "modus operandi." In this latter respect, I got to know Institute Prof. Arthur Von Hippel who was a pioneer on the dielectric properties of foods and their measurement.

Personal Matters

In 1962-1963, I served as the President of The New Century Club, after serving on the Board of Directors for several years. This organization was a professional club founded in 1900 by professional men of the Jewish Faith with the original purpose of good fellowship among physicians, engineers, dentists, attorneys and accountants. This was at a time when Jewish physicians had difficulty in being admitted to practice in the hospitals in Boston. Later, when the hospitals opened up to them, the raison d'etre became scholarship funds for needy students of the Jewish Faith. This was an active organization until about ten or fifteen years ago when there were too many activities competing for people's time, and scholarship moneys became more easy to obtain. It simply transferred its funds to another charitable organization in the area and went out of existence. It had been a great organization. I had a great time meeting many important people. One of the Honorary Members was Prof. Harry A. Wolfson, of Harvard, for whom we ran a great 75th birthday bash. Another notable speaker that I arranged for during my term in office was Lt. Gen. James J. Gavin, after he had served as Ambassador to Paris for Pres. Kennedy. We (Diana and I) really kept busy in those days!

FIG. 15.1. DIANA AND I GETTING READY TO STEP OUT TO A FORMAL FUNCTION, CIRCA 1962

During the period covered above, our three children grew and were thriving. We moved from Norman Road to 228 Upham St., purchasing the large Robie House at a bargain price. It had 5 bedrooms, 5 bathrooms, a large living room, a large informal parlor which we used as a TV room, a den (called the billiard room on the original blueprints of the house), a tremendous kitchen, a large dining room with mahogany wainscoting and a built-in leaded glass china cabinet, and a relatively small, octagonal breakfast nook. It also had a ballroom on the third floor and a built-in cedar storage room. The basement contained a tremendous power plant, gas fired.

We purchased the larger house because our parents were getting older and we knew that sooner or later one or both sets might have to come and live with us. In the event, my mother became senile and in 1961, had to be admitted to the Hebrew Rehabilitation Center, and my father, after he had suffered a heart attack in 1964, came to live with us. Diana insisted that he move in with us. She really took care of him, watched his diet, made sure the cabs came and took him to my mother's as scheduled and periodically, I would take him to visit my mother. Dr. Rovner used to come and visit my father and check him over carefully at regular intervals.

Errol was not doing as well in school as we had hoped and we transferred him to the Chauncy Hall Boys Preparatory School, then in Cambridge. He did well there and graduated in 1959 and entered Cornell destined for the College of Veterinary Medicine, which he said he wanted to do. He came back, not doing too badly, but had developed a lump in his right cheek which turned out to be a lymphoma although at first, on the surgical intervention, was originally thought to be a mixed tumor of the parotid gland. Dr. Richard Webster, the eminent plastic surgeon whom Dr. Rovner had asked to do the surgery, could not get all of the tumor out. So over a period of years, Errol and we suffered — through radiation treatments, etc. Errol did get married on April 1, 1962 and lived on Beacon St. with his wife, the former Linda Ward.

The final management of Errol's medical care was managed by the Leahy Group with treatments being done by Prof. Trump on Errol's spine, where metastases had developed. Dr. Trump's medical associate was Dr. Ferdinand Saltzman of the Leahy Group, and together they extended Errol's quality time on this earth some six months or so. Errol finally died on Friday night, May 24, 1963 at 8:45 PM at the Deaconess Hospital. It was a painful experience and Diana suffered more than I. It is horrible to be so helpless when a child is suffering and dying. But, as Rabbi Warren, who had married us and who officiated at Errol's funeral stated, we had done all we could. He was in God's hands. He was buried on Sunday, May 26, 1963 in Diana's parents' family plot in the Temple Beth El Cemetery in Chelmsford, Mass. We had over 150 people come to our house on one night during the brief shiva period (due to the Shavuoth Holiday). I still go each year to tend the gravesite, trim around the stone, say kaddish and have the El Mole Rahamin prayer said. May God rest his soul. He lies next to Diana's parents. However, life must still go on. Judy's confirmation at Temple Tifereth Israel, in Malden was May 29, 1963.

Judy went to Melrose High School, did extremely well in her studies and graduated number three in her class of 1965. She was admitted to Pembroke College in Brown University, and graduated from Brown with honors in the Classics in 1969.

Jonathan graduated from Melrose High School in 1971, after a lackluster formal educational career at M.H.S., but did he blossom in his last two years and, was awarded two scholarships at graduation — the Beethoven Society award for his musical ability and an award as an all around helpful student. He was a "big wig" in the Drama Club with his operatic tenor voice and participated in important roles in many of the Drama Club's shows. He too did us proud. He was also, as was Judy, a member of the Melrose High School Debate Club and participated in many debates. Diana attended most of them and ended up as chauffeur, and a judge at debate contests when one of the teams was not the M.H.S. Jonathan just hated the formal math, Latin, etc., classes. But later on, not to be left behind, on his own initiative, he received his M.S. in Management from Lesley College, going to graduate school nights while working at Polaroid.

Bar Mitzvah of Jonathan Mark Goldblith, July 4, 1966

I'm getting ahead of my story, however. On July 4, 1966, Jonathan was Bar Mitzvah at Temple Tifereth Israel and did a magnificent job. Rabbi Dr. Herman Pollack, our friend who was the Hillel Rabbi at M.I.T., gave the talk ("derasha") and Rabbi Judah B. Miller, our Rabbi, officiated. Diana's parents and my father were there as were many of our friends. Diana had on a special dress, which she had made in Hong Kong the year previously, and a matching hat which our friends, the Rosens of London, made for her. She looked gorgeous.

We held a large buffet luncheon, kosher catered, under tents in the side and back yards of our large house. It was a magnificent affair outdone only by the quality of Jonathan's

FIG. 15.2. JONATHAN MARK GOLDBLITH, JULY 4, 1966

FIG. 15.3. JONATHAN NEXT TO TEMPLE BULLETIN BOARD

FIG. 15.4. JUDY AND JONNY AT PARTY IN YARD, JULY 4

FIG. 15.5. JONNY AND DIANA WITH THE BAR MITZVAH CAKE

FIG. 15.6. AFTER THE PARTY, JULY 4, 1966. L TO R, MY FATHER, SAG, DIANA AND HER FATHER

"majestic" performance (to quote Rabbi Pollack's description). We had guests from the various generations of the family, close friends of ours from M.I.T., Lawrence, Lowell, etc. Sam Lepkovsky came with a case of "Chateau d'Yquem on his shoulder."

United Fruit and Food Connection

In 1961, I had been approached, and with President Strattons's agreement, had agreed to work part-time as a consultant but with a title of Vice President for the United Fruit and Food Corp., their new freeze drying and processed food subsidiary. There I got to know Mr. Peter Harris, Vice President of the parent company and General Manager of the UF&F Co. subsidiary.

One of the first things we had to do was become aware of what had been going on in the freeze dehydration industry in Europe and I was selected to go. In December, Diana and I took off for Europe — our first trip there.

The Beginning of The Peripatetic Goldbliths

Our first trip to Europe in early January, 1962, the first of many more to come, was for the purpose of visiting all the freeze dehydration establishments that we could plus the equipment manufacturers, as well as other pertinent visitations that I might feel worthwhile. I recall that we received much useful advice regarding this trip (as on many other occasions) from Ray and Geula Pariser, dear friends of ours who were there in the Department. The Parisers actually met in London during WWII when both were students — he at Cambridge in Chemistry and she in London studying Arabic. We have been close friends since 1957. We took a Pan Am nonstop to London, changed early the next morning for a BA Vicount for Edinburgh, were astonished to observe some bushy, walrus-mustachioed Britisher order ale with his full English breakfast. Diana and I could barely drink some diluted coffee and eat a biscuit. At Edinburgh we transferred to a DC3 aircraft bound for Aberdeen. We were scheduled to visit the Experimental Factory (freeze drying) of the Ministry of Foods and to see its director, Mr. Sidney W.F. Hanson, who, we learned later, was winging his way across the Atlantic enroute for Chicago, where he had "a command performance" at Armour and Co., who had, if memory serves me correctly, just purchased the Experimental Factory. It was Sidney's birthday, but the party went on without Sidney and we spent a warm (internally) cold (externally) night at the party and thereafter at the Station Hotel in Aberdeen. I saw the Experimental Factory the next day.

We then went back by plane to Edinburgh and visited our friends the Greenwoods. Alan was the founding Director of the Poultry Research Center in Edinburgh. Diana bought some sheep-lined overshoes. Her feet were freezing!

We then went to London where I visited Vickers who made lyophilizers. We stayed at the Cumberland Hotel at Marble Arch and visited with Miss Mamie Olliver, the eminent British Food Technologist, who worked for Chivers as Chief Chemist. I also visited J. Lyons and Co. and had lunch at the Cafe Royal (which they owned). Later we flew to Amsterdam and visited the headquarters of the European United Fruit Co. — Elders and Fyffes — in Den Hague. Then we took the train to Cologne (Koln) where we saw the Dom Cathedral and I visited Leybold Hochvakuum (freeze dehydration equipment manufacturers). Next we took the train for Karlsruhe where I visited with Prof. John Kuprianoff, whom we had entertained at our home at one of the conferences I have mentioned earlier. He took us to a lovely restaurant in Ettlingen in the Black Forest where we had a delightful meal. John was a great linguist, as well as a first rate engineer and his laboratories were outstanding. He was a true "renaissance man" with knowledge of and a collection of Oriental carpets and Japanese wood block prints.

From there we went to Zurich where I visited the ETH and met also Prof. Walter Pilnik and his wife, Rena, with whom we were to become close friends to this day. Pilnik was the Managing Director of "Obipektin," a Cooperative in Bischoffzell, near St. Gallen. He manufactured some very tasty strawberry powders made by "puff drying," in which the UF&F Co. became interested shortly thereafter. Then we visited Knorr who freeze dried many of the vegetables for their soups. There I also saw two former students. The next day, we went to Vevey, by train, to visit with the Nestle Co. I also gave a lecture at Nestle and discussed freeze drying and irradiation preservation of foods with them for two days. We were treated royally there. I must have returned home twenty pounds heavier. Three weeks after the trip began we were back home. It was tiring but eye opening. The Europeans were, to my mind, well ahead of us in a number of aspects of lyophilization.

The Grand Tour of Europe in Summer of 1962

Our second trip to Europe was in 1962 and we took Judith, Jonathan and Judy's best friend, Judith Lindahl. It was "the Grand Tour," or as much of a grand tour as I could make of it inasmuch as I had some business to do for the United Fruit and Food Corp., as well as to meet researchers and faculty members in similar fields to those in which I was interested, and in which I was working at M.I.T.

We met Robert (Bob) Gillis, a young marketing executive that Pete Harris had hired, who was put in charge of the UF&F Co.'s European lyophilized shrimp marketing effort. We met in London, as I recall it in the summer of 1962, where we were staying at the Hotel Goring. Judy Lindahl was, like Judy Goldblith, age fifteen. Jonny was age nine. Judy L. is now an attorney of some standing in Boston, specializing in criminal law.

Our stay in London was in two segments, before and after a trip to Grantham, etc., as described further on. In London, while I did business, Diana and the three youngsters did the Houses of Parliament, Westminster Abbey, the Old Curiosity Shop, St. Paul's Cathedral, etc. After three days or so, I rented a car and we drove to Grantham inasmuch as Judy wanted to see the Inn at Grantham mentioned in, I believe, Nicholas Nickelby. Judy G. was then quite a fan of, and well-read in general, but especially of Charles Dickens. We had been staying, (before we returned to London

FIG. 15.7. BEGINNING "THE GRAND TOUR," JULY 1962, DIANA, JUDY GOLDBLITH, JONNY, JUDY LINDAHL

where we stayed at the Goring), at the Hotel Park Lane when we took off, oh so slowly, with a flat tire to begin with, for Grantham.

After an overnight at Cambridge, where I visited the Low Temperature Station (LTRS) of the Ministry of Foods, which was tied in with the university, and Diana and the children were "entertained" by the wife of the director, Mrs. E.C. Bate-Smith (her father had been the founder of the LTRS, Sir William Bate Hardy. Diana and the children toured the Fitzwilliam Museum under the guidance of Mamie Olliver (who lived at Histon, nearby), as well as other parts of that wonderful city of the other Cambridge. We spent one night there at a small hotel and then left for Grantham where we stayed overnight at the Inn at Grantham. The next day we began tc drive to Stratford on Avon and saw a play that evening which I believe was either Cymbaline or Corianus.

We then returned to London to the Goring where we met Bob Gillis and then flew to Amsterdam where we did Den Hague, Amsterdam and Rotterdam in a few days. In the meantime, I received a telephone call from Walter Pilnik who asked me to come to Zurich for a day to be the "Sandek" (Godfather) for his younger son who had been born just a few days earlier. So I left Diana and the gang in Den Hague and I flew in for the Brith Milah of Udi Pilnik. It was quite an experience. I flew back that evening to Amsterdam and caught up with Diana and the youngsters and on we went.

The next day we flew from Amsterdam to Paris and stayed at the Trianon Palace at Versailles. We had a two day tour of Paris in a car with a driver named "Einstein" and saw the Eiffel Tower, the Louvre, took in the Follies Bergere, etc.

From Paris, in a car I had rented, we went to Lyon via Dijon and the wine country. Diana and I enjoyed it. I'm not sure that the youngsters did — eating lunch outdoors under the trees, etc. We stayed several nights in Lyons at the Hotel Royal, inasmuch as I had to lecture at a lyophilization course in Lyons run by Prof. Louis Rey, then a Professor at the University of Dijon, and later head of research for the Nestle Co. One of the side trips which was part of the Course was a trip to the medieval Citie of Perouges, near Lyons, where the inn set a superb table and the local rose wine, "Cerdon," was outstanding.

Then we drove to Geneva and Vevey where I had to be with Bob Gillis on Nestle/UF&F Co. business, as well as to give another lecture. Diana thought it would be nice to drive to Zurich over the mountains through Interlaken. It was quite a drive in the non-air-conditioned, non-power-assisted steering, large Citroen car of that era. But we made it OK and everyone was happy. I believe that was the nub of the Grand Tour of 1962.

In 1963, I found an invitation to attend the Fourth Japan Atomic Energy Industrial Forum to which Diana and I went later that year, visiting not only Kyoto, but Tokyo, where we met the entire Ayukawa family and reported about Junta, Masako and

Yaichi, and also met Masako's father (her mother was ill on that day). Her father was the Grand Chamberlain to the Emperor. We also got a chance to see Dr. Tei Yaminishi, who had been a post doc in the Department circa 1957, and a number of former graduate students and other post docs.

FIG. 15.8. DR. TEI YAMINISHI, OCHANIMIZU UNIV. Professor, Diana and SAG, in Kyoto in early 1960's.

After Japan, we went on to the Philippines where I visited with Sr. Marie Arthur, and it afforded me an opportunity once again to see my classmate Oscar B. Mapua and his wife who had also helped to save my life by sending money to me at Cabanatuan via the caribao cart driver. Both were emotional experiences and I believe that I have covered them earlier.

The meeting with Sr. M. Arthur was most moving. We were delighted to meet once again after the turbulent war years, and I was so happy to be able to have Diana meet Sr. M. Arthur.

While in Manila, Dra. Rosalinda Valenzuela, who had found Sr. Marie Arthur for us in Quezon City, had arranged for us to go to Baguio and came with us as our guest. We met Sr. Sussanah, who had been in Bontoc during the war and Diana had a chance to see Baguio as did I, under more normal circumstances. We flew both ways. Dra. Valenzuela arranged for us to go with the staff of the Public Health Dept. on their official boat to Corregidor. It was a thrill for me especially and I am sure for Diana. I photographed every gun emplacement and battery, etc. What a day! I also spoke that week to the M.I.T. Club of Manila.

FIG. 15.9. DIANA, SAG AND SR. MARIE ARTHUR AT ST. THERESA'S COLLEGE, QUEZON CITY, 1962

The next day we took off for Bangkok via Saigon. Armed guards were then in the airport. We stayed two or three days in Bangkok and saw the Cheosakuls and went on the "floating market," as well as to a number of Temples. I spoke at a luncheon meeting to the M.I.T. Club of Bangkok. Then we took off for Israel via New Delhi and Teheran. It was our first visit to Israel and it was glorious. We went on the Friday Synagogue tour, we went to Capernaum, Haifa, where I met a cousin, Israel Lerner, Tiberias, etc. I am sure that we returned home via Zurich, where I had some business with Dr. Pilnik.

In 1964, I spent all of July alone in Europe, where I mainly was lecturing at a Lyophilization course in Lyons, France conducted by Prof. Louis Rey. At the end of the course, I went with Prof. Arnold Spicer of Ranks, Hovis and MacDougall to Paris and on to Rome and Salerno, where he wanted to look at a freeze dehydration plant owned by Sperti Industries. George Sperti was a creative individual, whom I had met after I became a member of the Board of The Franklin Corp., an SBIC, in 1964, I believe. Sperti (the inventor of the Sperti Miners' Lamp) was also President of the Institutum Divi Thomae, where John Loofbourow, Prof. of Biophysics, had been before he came to M.I.T. Sperti had invested in an Italian Company in Salerno, which was in the freeze drying business, and they were trying to sell off the company to Arnold Spicer, as I recall it. It was not much of a company with old-technology Leybold equipment. It did give me a chance to meet an Italian Count, be driven at ultrahigh speeds from Rome to Salerno, saying prayers all the time in Hebrew, eat an Italian meal at Salerno and rush back to Rome for the plane to Amsterdam, where I met the Pilniks who had just moved to Wageningen, where Walter had become head of the Agricultural University Food Technology Institute.

The trip gave me an opportunity to meet one of Sperti's investors, a New York lawyer, named William Shea (whom I saw again in New York at his office) for whom the Shea Stadium in New York is named. Then I went to Vienna for an "expert" meeting on Food Irradiation at Vienna and then home via Paris. I had been gone a long time alone — too long, and vowed not to do that again without Diana.

Around the World a Second Time (Nov. - Dec., 1964)

In late 1964, I had an invitation to attend a Japan New Food Business organization meeting and to address them in Kyoto and deliver some lectures, elsewhere, as I recall it. Beginning on November 8th, Diana, Jonathan and I took off for California where I visited some linear accelerator facilities and manufacturers in the Bay Area, Lep and other friends there, then flew to Los Angeles and on to Honolulu for a couple of hours and then we departed for Tokyo. We spent a few days with Mrs. Ayukawa in Tokyo (Yaichi was still in the Boston area), and then left for Manila via Hong Kong. I had the pleasure of introducing Diana once again and Jonathan for the first time to the Mapuas and to Sr. Marie Arthur.

On Saturday, November 22, we went, courtesy of an M.I.T. Alumnus in the Philippine Ministry of Health, to Bataan by large station wagon. We could only get as far as just north of Cabcaben, but at least I had brought some of my family to Bataan. We then went to Israel via Hong Kong. In Israel, I met with the leader of the food irradiation program at the Israel Atomic Energy facility, Dr. Lapidot, took Jonathan and Diana from Rosh Pina in the north (by plane) to Eilat in the south. There we met Leon Eligoulachvili, a Parisian investor whom I knew, to the surprise of Diana and Dr. Lep, who had met up with us a few days earlier in Tel Aviv.

On December 5th, our friends the Mendes Sachs (he had been a student of Lep's years back, was then the chairman of the Citrus Marketing Board of Israel and later of the First National Bank of Israel) took us to dinner and to a concert of the Israel Philharmonic at Mann Auditorium. Another guest of Mendes Sachs was Mr. Sol Wishniak of Mexico, who was a customer of Mendes' and whose name sounded familiar, so I asked him if he was related to my student, Herbert Wishniak Weinstein of Mexico City. He replied in astonishment that Herbert was his nephew. It's again a small world!

From Israel we flew to Vienna via Frankfurt, went to the opera, stayed at the Sacher Hotel and had some Sacher torte. I did some IAEA food irradiation business there with Dr. Harry Goresline, formerly of the Food and Container Institute for the Armed Forces in Chicago. The next stop was Copenhagen for a day or two, and thence to Gothenburg, Sweden, where I was to visit with Prof. Erik Von Sydow, the Director of the Swedish Food Preservation Institute, and to deliver a seminar there.

FIG. 15.10. DIANA AND SAG JUST PRIOR TO GOING ABOARD LAKE BOAT FOR TOUR OF UPPER PART OF LAC LEMAN.
Picture taken at dock at La Tour De Peilz, Circa 1964.

The first item of business the next day was the opportunity to participate in their Yuletide Santa Lucia festival. We were all thrilled, especially Jonathan. It was the only time that we have ever seen a Santa Lucia festival, firsthand. Erik and his wife have become old and dear friends since those early days. (I should have mentioned that Erik arranged for us to meet and have dinner with his elder brother, while we were in Hong Kong. They are both first cousins of the famous Swedish actor and movie director, Max Von Sydow).

From Gothenburg, we flew to London, had dinner at the Spicers. The Rosens were there and drove us back to Brown's Hotel. The next day, we left, I for Cambridge for a day at the LTTR, at their request and Diana and Jonny stayed at The Bull at Garrards' Cross. We three then went to Bedford where we stayed at The Bridge Hotel (heat supplied by a small smelly gas heater which ran on shillings, plunked in). I visited the Unilever Laboratories in Colworth House at Sharnbrook, nearby, for a day on December 22nd. We left London on December 24th although delayed by foggy weather, we took off several hours later for New York. But, the weather was bad at both JFK and in Boston, so Pan Am landed in Montreal, where we were put up at a second-rate hotel and arrived in Boston, via Pan Am on December 25th, 1964. It was a long hard trip. Mrs. Blake stayed with Judy and my father, (Zadie). All of us were happy that we were home.

We had traveled many miles in 1964, made new friends and revisited old ones. Our children had an opportunity to see some of the old world, firsthand and Jonathan had been to the land of his forefathers, as well as traveling around the world.

In the summer of 1965, I had some business in Europe, mainly to do with RHM and Nestle, and brought Judy with me as a pre-college, post-graduation treat. It was good for both of us. We also visited Mr. Henri Cheftel, the next Underwood Prescott Lecturer in Paris.

In 1967, I made another trip to Japan for an Atomic Energy Industrial Forum Meeting with Diana. It was after the Six-Day,

Arab-Israeli War, and Diana and I returned to the United States via the Philippines and Israel. We saw the Golan Heights from atop and overlooked Tiberias on one side and Syria from the other. My second cousin, Oded Lerner, a graduate of the Technion, led a tank group up the hills and told us all about it. It was on one of those trips during the sixties, (not on that trip) that we brought Pat Proctor with us to Japan. It was an eye-opener for her to see the Far East and its culture.

Other Family Events During the Decade of the 1960's

The decade of the 1960's represented life as it really is, for my family and me — truly life-cycle events. In 1967, both my parents passed away. My father had been living with us on 228 Upham Street since 1963, when he had his first heart attack in Lawrence. Diana prevailed upon him to come and live with us. He

FIG. 15 11. JUDY AND HER "ZADIE," MY FATHER, IN MID 1960'S

was a bright and pleasant individual, with all his faculties. Gerry Rovner looked after him medically. When Diana and I went away, we were able to obtain live-in help from one of the aunts of my secretary, Janet Anderson, Mrs. Blake. She was good for both Johnny and my father. When we went out for an evening, we don't know what happened. Did the two youngsters look out for my father or vice versa. But, he never complained nor did they.

Death of Abraham Goldblith and Fannie Goldblith

Pa wasn't feeling too well on the evening of July 31st, 1967 and passed away, early the next morning at the Melrose-Wakefield Hospital, where Dr. Sidney Levine, another friend, took care of him as soon as he arrived, by ambulance, at the hospital. It was the second heart attack he had while in Melrose. He passed away peacefully at 3:00 AM on the morning of August 1, 1967. The funeral services were conducted at the Congregation Tifereth Anshei Sfard on Lowell St. in Lawrence, and he was buried in Lawrence at the cemetery of the Congregation Tifereth Anshei Sfard in South Lawrence.

My mother passed away at the Hebrew Rehabilitation Center for the Aged in Roslindale on Friday, December 1, 1967 at 4:45 PM. The funeral was at the same synagogue as my father's, of which both had been lifelong members, on December 3, 1967.

In the same year, I had been made deputy department head.

In 1968, after we had decided that we no longer needed such a large house, Diana began to look around for a more suitable size residence for us. Judy would be soon getting on with her life and Johnny was not far from getting through high school and off to college. Diana found the house in which we have continuously resided since 1968, when we sold the 228 Upham St. residence, and moved to Meadowview Rd. Since 1967, we have been helped immensely by our good friend Mrs. Sophia Tammarro who helped Diana on Upham St. in the immense house, and since then here on Meadowview Rd. She continues to help me part of two days per week, Bless her.

Marriage of Judith Ann Goldblith to Richard Douglas Clark

In 1967, Judy met a very nice young man at Brown who was to become her husband, in 1969, Richard D. Clark. Diana and I were both delighted. Rich graduated in 1968 with an A.B. in American Civilization and became a school teacher in Aurora, Illinois. Judy graduated from Brown with an Honors degree in the Classics, on June 2, 1969. They were married on July 27, 1969 at

FIG. 15.12. MR. AND MRS. RICHARD D. CLARK AT THEIR WEDDING, AT TEMPLE TIFERETH ISRAEL, MALDEN, MASS., JULY 27, 1969

FIG. 15.13. DIANA AND I
AT JUDY AND RICH'S WEDDING

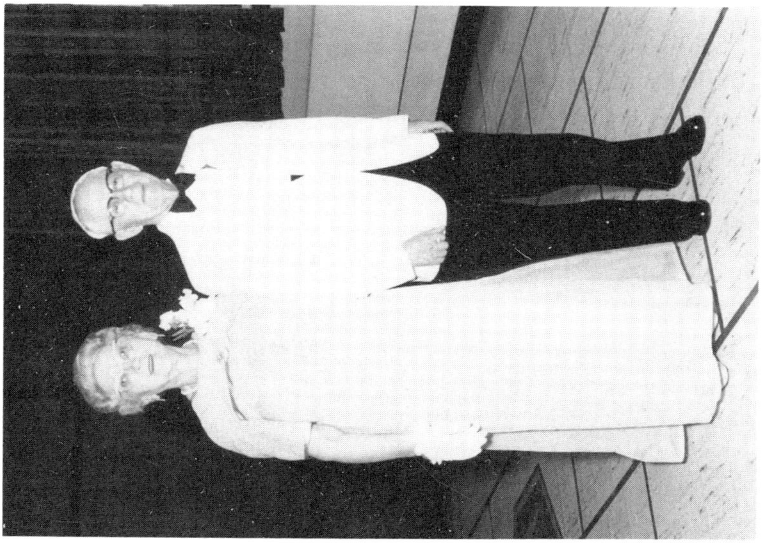

FIG. 15.14. DIANA'S PARENTS,
IDA AND GEORGE GREENBERG
AT JUDY AND RICH'S WEDDING

Temple Tifereth Israel in Malden where Judy had been confirmed. It was a lovely wedding and a happy one. Rich's parents were grand and his mother is still living in Geneva, Illinois. We met a great many of Rich's family who came for the wedding. The wedding was officiated by Rabbi Dr. Herman Pollack and Cantor Sheldon Chandler. (Rabbi Miller was away at the time). Herman was a great old friend, who sadly is gone now as is his wife.

FIG. 15.15. DEAR FRIENDS OF OURS AT JUDY AND RICH'S WEDDING, SAMUEL ISSENBERG, RUTH AND MILTON ISSENBERG

FIG. 15.16. CATCHING JUDY'S GARTER; UNCLE MAX GREENBERG AND HIS GREAT NEPHEW, JONATHAN GOLDBLITH

Birth of Sarah Beth Clark

Rich taught school in Aurora, Illinois for a few years after they were married and Judy worked in social work for the State of Illinois at Aurora. Then they decided to begin a family and Judy delivered a lovely baby girl on May 10, 1971, whom they named Sarah Beth Clark. I happened to be on consulting business at Armour and Co., in Oakbrook, Illinois when she developed a complication, while in labor, and would require a Cesarean Section. Rich got hold of me by telephone and I was able to see Judy before she was sedated for the C-section and then again after Sarah was born. Both looked great and I could return later that night to Melrose and tell Diana all that I knew about it and that I had seen our first granddaughter. Diana came out for a couple of weeks to give Judy a hand after she got out of the hospital. Those were the days before turn-about births. I might mention that this was my first trip in several months inasmuch as, in February, 1971, following a one week's trip to Japan, I developed thrombophlebitis and had to be hospitalized for a couple of weeks.

I celebrated my fiftieth birthday on May 5, 1969 and a group of M.I.T. friends threw a lovely party for me on May 3rd, at Les Tuileres Restaurant at the Hotel Eliot in Boston, which was a favorite of ours. They presented me with a silver bowl which I still treasure.

The years 1969 and 1970 brought forth some more pleasant news for us. In 1969, I received the Distinguished Food Scientist Award of the New York Section of the Institute of Food Technologists in March of 1969 and the Babcock Hart Award for Nutrition from the I.F.T. at its Annual Meeting in Chicago in June of that year. In 1970, I received the Nicholas Appert Award of the I.F.T. in San Francisco, at the Annual Meeting. This was a momentous occasion inasmuch as, at the same award dinner, Marc Karel received the first William Cruess Award for excellence in teaching and Steve Tannenbaum received the Samuel Cate Prescott Award presented to a young (under 35 years of age) member who has done outstanding research. We all had a "bang-up time" that evening.

The Clarks Move To Melrose

The most pleasant news of that period in our lives was Jonathan's graduation from Melrose High with two awards (as related elsewhere herein), and his admission to Emerson College to study Theater Education *plus* Judy and Rich's decision to move to Melrose in 1973.

Rich found a job as a teacher in the Melrose School System for which he was well-qualified. He already had a M.Ed. degree from Northern Illinois Univ. at DeKalb plus five years experience. We were delighted, of course. In due time Rich began to work in the business field, having worked one summer at the William Underwood Co. He remained on at the Underwood Co. for a number of years, in positions of increasing responsibility.

Birth of Second Grandchild, Rachel Deborah Clark

On July 25, 1974, our second granddaughter, Rachel Deborah Clark, was born. We were, of course delighted, to say the least. She was a pretty baby who looked just like Judy. She has had all the duties, etc. as the baby of the family — especially the Asking of the Four Questions every Passover Seder. More about "Rachie" later.

Death of George Greenberg, Diana's Father

The winter of 1975-1976 was a difficult one in terms of a flu virus. The virus hit the nursing and retirement home, Willow Manor, where Diana's parents were living and her father came down with the flu and couldn't "beat it." He passed away on March 1, 1976 and was buried in Chelmsford the next day at the Temple Beth El Cemetery. Jonathan came from New York to be with us. George was buried on a snowy Tuesday by Rabbi Lebeau of Temple Beth El, which Ida and George helped to found and where we were married decades earlier. He was a great human being, thoughtful, kind and helpful. We missed him very much as did Diana and her mother. Grandma Ida insisted and came to the cemetery, snow and all. She survived him by over seven years. He followed politics avidly and helped to lead the Republican Party in Lowell. I still miss Ida and him as I do my own parents.

Friendship with the Rosens of London and their "Mischpocha"

During the 1960's, at a dinner party at the Spicers who lived then in Iver Heath, Bucks, we met two of their friends with whom we became close friends. Through them we met the Rosen's entire family as well as having the pleasure in the early 1970's of spending two winter holidays with them, one in Ochios Rios, Jamaica and another in Tobago. Dolly Rosen passed away early in this decade but Gary Rosen is still, touch wood, hale and hearty and we correspond regularly. They designed and made the hat that Diana wore at Jonathan's Bar Mitzvah in 1966.

Through the Rosens, I also remet Marjorie Ostrer who had worked with Lep at Berkeley, years earlier and her lovely daughter, Isabella, and later the man she married, Trevor Blench, all of whom, we, or I alone, saw from time to time in Switzerland where they live in Geneva. Trevor's father had worked for the Nestle Co. It's once again, a small world. Diana and I became friendly with Donald Walker who was married to the Rosen's, daughter, Della. They broke up after some years. We also saw and became friendly with their daughter Pamela and her husband Ted Ostrer, nephew of Isidore and Marjorie Ostrer (Lep's former student). Diana and I used to visit with Donald Walker, and I with Donald's second wife, Irene, once in awhile. I am fond of them both and of the Walker children, who are now all successful in their careers.

Other Public Service Activities in the 1960's-early 1970's

As a result of President Kennedy's meeting in the Rose Garden of the White House with the then president of Argentina, President Frondisi, a Joint U.S.-Argentina Commission on Foot and Mouth Disease was established. The first chairman was Dr. J. George Harrar, the then president of the Rockefeller Foundation.

The U.S. members were mainly veterinary virologists. I was the only food technologist of the group. The Argentinean counterpart was a distinguished group including the then only South American Nobel Laureate, Dr. Bernardo Houssay and the future Nobel laureate, Dr. Luis Le Loir. The mission and its members were considered sufficiently important, so that a junior physician, attached to the White House Science Advisor's staff, was sent along with us on our first visit to Argentina in early 1962. We visited a number of slaughter houses, meat packing plants, and a cattle ranch where we were treated to quite a "gaucho" rodeo and feast on this our first trip.

We stayed at the Hotel Plaza and I met some of my Argentinean first and second cousins, descended from my mother's eldest brother who had emigrated from Poland near the turn of the century. He settled at the western border of the Buenos Aires

Province. It would have taken me a day by train (24 hours) to visit him or a somewhat shorter time by plane. But both schedules were not very dependable. So I had to content myself with visiting with his son, Pedro Rubin, my first cousin and some of his family. My mother who was then alive had given me their address. Uncle Leon was then, in 1962, in his early nineties. I had considered the private plane possibility, but having flown in one to Cordoba, which was held together by "bailing wire," I decided against taking the chance. (The Cordoba trip with the Joint Commission members in a private relatively-large, twin-engine, piston-driven propeller plane was quite an experience).

Some time after the first mission, the White House transferred the Argentina-U.S. Commission to the N.A.S.-N.R.C. Needless to say, the problem of Foot and Mouth Disease is still a major animal disease problem. However, major advances have been made in new vaccines. Food Technology offered but little promise in developing raw meat products free of the virus.

I made about three more trips to Argentina and Chile and two to Rio de Janeiro, where the Pan American Foot and Mouth Disease Laboratories were located. They were under the directorship of Dr. William M. Henderson which whom I became very friendly. My ulcer was acting up at that time and Greg's wife Beryl made sure that I had boiled milk to drink while I stayed with them. Greg Henderson deserves a great deal of credit for whatever success the meetings had. He performed an admirable service, was knowledgeable in not only the virus and the scientific aspects of the problem, but was trusted by both the American and Argentinean groups. At our opening session, we were greeted by Dr. Houssay who stated that "Aftosa was not a medical/scientific disease but rather a political disease." What a way to begin!

My friendship with the Hendersons continues to this day and we have visited each other a number of times in the past 30 years. He was the second to last Underwood Memorial Lecturer, and spoke at the Symposia associated with the Paul E. Gray Inauguration as President of M.I.T. in 1980.

Another public service involvement of mine was that of being a member of a select panel on The World Food Problem appointed by the Office of the White House Science Advisor in 1965. Again, Dr. Harrar was chairman of this panel. I served as a member of the main panel and as chairman of the sub-panel on protein. We produced a splendid report and this was a very satisfactory experience for me in meeting leaders in the fields of food business, technology and nutrition.

The UF&F Co. venture into processed foods did not work out well and several of us who came in near the beginning of the business resigned and went on to other work, which for me was consulting, as well as at M.I.T., where I was still a faculty member and executive officer of the department.

I began to consult for Armour and Co., General Mills and The Stop and Shop Cos. (the latter lasted until the early 1990's), and the Stauffer Chemical Co., which later established a scholarship at M.I.T. in my name in the early 1970's for some special work I did for them. I have had pleasure in contributing more to that scholarship fund since that time.

My research activities grew and expanded more into the field of microwave processing of foods and lyophilization. Although, in the field of Radiation Preservation of Foods, I chaired the planning group for a major symposium on this topic sponsored by the USAEC and the Natick Laboratories of the QMC and the Advisory Board on Military and Personnel Supplies of the N.A.S.-N.R.C. This took place on September 27-30, 1964 in Boston, and had speakers from all over the world. I noted in the opening address, which I made, the progress that had been achieved, as well as what was needed for the future.

Additionally in this period, I became a consultant to the Nestle Co., and went to Europe to Vevey at least once or twice per year. This resulted initially because of my lecturing at the International Courses on Lyophilization organized by Prof. Louis Rey who came to Nestle from Dijon University as Manager of Research.

As is obvious from the above, I frequently began to go to England in the mid-1960's to visit laboratories in fields in which we had mutual interest and became friendly with Prof. Arnold Spicer, a food executive who sold his business to Ranks, Hovis and MacDougall, and set up the large Lord Rank Center for Research. I have mentioned him earlier above. Arnold had attended a course which we presented in freeze drying and irradiation in the mid-1960's and we became friendly. Through Arnold, I met the Lord Rank (J. Arthur Rank), Sir Peter Reynolds, the Managing Director of the Co., and his lovely wife, Anne, who was very nice to Diana on a number of occasions. Later, I remained as a consultant for the company after Arnold Spicer retired, to be succeeded by Prof. Jack Edelman. When the addition to the Lord Rank Research Center was dedicated, I was invited and met the Queen Mother, Queen Elizabeth, who dedicated the building in the mid-1970's.

In 1962, I had met Lord Rank who was the host, at the First International Conference of Food Science and Technology in London, which Marc Karel and I attended. Lep came over from Holland and we rented tails at Moss Brothers emporium to attend the formal Lord Mayor's Banquets at Guildhall. It was quite an event.

I got to know Lord Rank quite well in the 1960's and 1970's. He, for some reason, always referred to me as "Colonel Goldblith." That was J. Arthur Rank!

Death of Dean Prescott

I had been close to Dean Prescott during his latter years, when his office adjoined mine. From him, I learned all about his relationship with William Lyman Underwood, about their early work together in Maine, etc. Frequently, I brought him to his M.I.T. office if the weather was bad, and whenever possible returned him to 100 Memorial Drive where he lived for about the last decade of his life. I gained a lot thereby. We had one I.F.T. Northeast Section meeting at the Underwood Co., and it was a pleasure to see him beaming as George Seybolt talked about his

FIG. 15.17. SAM (SR) LEPKOVSKY AND SAM (JR) SAG BEFORE THE LORD MAYOR'S DINNER AT GUILDHALL, LONDON, 1ST INTERNAT. CONGRESS OF FOOD SCIENCE AND TECHNOLOGY, 1962

and Lyman Underwood's work. George really loved and respected the Dean.

On March 19, 1962, he passed away due to a stroke — just a few weeks short of his ninetieth birthday, loved and revered by all who knew him and considered by most as the "Founder of the Institute of Food Technologists."

Establishment of the Underwood Prescott Memorial Lectureship

Many of us in the department and indeed in the Institute Administration thought it would be good if something were done to honor his and Underwood's pioneering work. Accordingly, I made a suggestion to George Seybolt, whom I knew to be a person capable of making a decision if he wanted to, that we might establish an annual lectureship in the Underwood-Prescott name. He liked the idea and so the Underwood Prescott Memorial Lectureship came into being shortly after Prescott's death.

Underwood-Prescott Memorial Lectures

The first lecture was presented by Dr. C. Olin Ball in September, 1963. It was held in the Little Theater of Kresge Auditorium preceded by a remarkable dinner at the Algonquin Club made from all canned foods. Over the twenty years that the award was given, the dinners changed to luncheons preceding, and the topics changed from legislation, to nutrition, to toxicology, to enzyme catalysis. But they all had a common thread — that of excellence — excellence in the stature of the awardee and of the program which expanded into symposia as time went on. We always had a senior officer of the Institute present to make the award and for the private dinner afterwards for the M.I.T. Committee, the senior officer from M.I.T., George Seybolt and a number of his Trustees — all with wives. In this way, we came to know the Paul Grays very well. They attended a number of the dinners that George Seybolt ran and Paul Gray presented a number of the awards as Chancellor. Later, of course, I became friendly with Paul as a result of being Director of Industrial Liaison and still later as Vice President.

The awardees were selected by a faculty committee from M.I.T. of Marcus Karel, John T.R. Nickerson, with SAG as Chairman. When Nick retired, Gerry Wogan replaced him. The last U-P Awardee was Prof. Ephraim Katchalsky-Katzir, Prof. of Biotechnology at the Weizmann Institute of Science and at the

University of Tel Aviv, and who had been the Fifth President of the State of Israel. I made many new friends through the U-P Lectures. I recall taking our daughter, Judy, to Europe, the summer after she had graduated from high school and was to enter Brown University in the fall. We visited, among others, M. Henri Cheftel,

FIG. 15.18. SISTER JEAN DELBECKE (MARIE ARTHUR) TAKEN IN MID 1970'S DURING ONE OF MY VISITS TO HER AT ST. THERESA'S COLLEGE

FIG. 15.19. SISTER JEAN DELBECKE AND SAG TAKEN ON THE SAME VISIT

who had been selected to receive the U-P Award that fall. The lecture series came to an end when the Underwood Company was sold to IC Industries and its food division, Pet Inc., did not wish to continue the sponsorship beyond the 1982 award.

Establishment of the Underwood-Prescott Professorship at M.I.T.

In 1972, following a presentation by then Vice President of M.I.T., Mr. Vincent A. Fulmer, and SAG, to Mr. Seybolt, the Underwood Company established the Underwood-Prescott chair endowed in perpetuity, provided the Institute would raise half the amount, to which the Institute agreed. The year 1972 was celebrated by the company as the 150th anniversary of its founding during the week of May 31, 1972. It was quite an event. One of the features was the announcement by M.I.T. on May 31, 1972, of the first holder of the chair which was to be, yours truly, SAG. I was surprised when I learned about it from a telephone call from Diana who had seen it in the morning *Globe*. Just as she had finished reading the announcement to me, Walter Rosenblith, the Provost of M.I.T., walked in my office door to inform me officially. He had meant to do it a day or two earlier but got swamped with work that also had to be done. I was dumbfounded and had not expected to be named inasmuch as I had been out on the road raising part of the funds that was to be M.I.T.'s portion. I really was embarrassed at first, but then the significance of the honor of being selected to have my name associated with the Dean, whom I knew so well, overcame the embarrassment.

My close relationship with the Underwood Co. and George Seybolt continued. He sought my advice about the purchase of Accent International and about other matters, as well. He appointed me a Board member of Accent for a five-year period. We remained close personal friends even after his departure from the company and until his death. I remained close to Jim Wells who succeeded George as President and CEO. It was a wonderful relationship that Diana, "Didi," Mrs. Seybolt, and I enjoyed for many years.

Friendship with James B. Lampert and the Industrial Liaison Program

Lt. Gen. James J. Lampert (U.S.A. Ret.) joined M.I.T. in 1972 as Vice President for Resource Development. We were introduced by Dr. Kenneth A. Wadleigh, a faculty friend for many years. Jim and I had much in common although he was a few years older than I. He had been the Executive Officer of the sister Aviation Engineering Battalion and he and his battalion were sent to Australia. Later, he became Superintendent of West Point Military Academy and retired from the Army in 1972. Of all the job opportunities, he had, he selected M.I.T. as his post-military position. He was a great person.

In the course of our talks at a number of luncheons, I told him about Yaichi Ayukawa. Of course, he knew the family name inasmuch as Jim's position previous to retirement was as High Commissioner of Okinawa just before its reversion to Japan.

Among Ken Wadleigh, Jim Lampert and Al Hill, the then M.I.T. Vice President for Research and my old Electronics subject professor, they talked me into helping the Industrial Liaison Program. I agreed to help them by bringing in some companies from the food industry as members of the program, and we set up a special evening dinner and a symposium the next day for CEO's of selected food companies by a blue ribbon group of M.I.T. faculty. It was a success and we signed up five new members in a month or two. This success led to more responsibilities, of a different nature, which will be discussed in the next chapter.

CHAPTER 16

DIRECTOR OF INDUSTRIAL LIAISON

On September 1, 1974, I was appointed Director of Industrial Liaison supposedly on a half-time basis, and retained the title of Underwood-Prescott Professor of Food Science. Marc Karel took over my duties as Associate Department Head, a title I received two years earlier together with increased departmental responsibilities. I told Prof. Nevin Scrimshaw, Head of the Department of Nutrition and Food Science, of my impending change at M.I.T. while we were together in New Orleans at a National I.F.T. meeting in June, and when asked about a suggested replacement, I suggested Marc Karel as the only one who could satisfy both major groups of the department and do an excellent job. Marc was appointed and did an admirable job.

The Directorship of Industrial Liaison involved a deep commitment not only of time but of spirit, creativity and imagination. It would involve creating loyalties from the staff that existed there at that time and would involve some major changes in policy. Instead of remaining as an organization responding to new companies brought in by senior officers as a part of a capital campaign, it would mean our going out and recruiting new member companies, servicing them all well, and getting more help from the faculty both in terms of seeing visitors and in the recruitment of new companies. This was to mean much more traveling, as well as a change in emphasis of what I did for M.I.T.

I moved right into the home of the I.L.P. in Bldg. 39. I cut the overhead by reducing the numbers of secretaries. I wanted the Industrial Liaison Program Officers out on the road servicing their member companies and seeking new members in their fields of expertise. Thus one secretary for two officers was sufficient. I added half again as much space for each I.L.O. for dignity, as well as a morale booster. I contacted faculty, offered faculty up to one thousand dollars in revenue sharing points for each new company that they identified or brought in. We received the help of the faculty in the I.L.P. as we never had before.

The revenue sharing system, the idea of Ken Brock's which I put into effect, helped a great deal. Ten percent of the entire gross revenue of the I.L.P. was available for the faculty in accordance with the help they offered the I.L.P., such as receiving visitors, visiting member companies, providing manuscripts for us to send to member companies, etc. We began to create some excitement. I led by example. I traveled to prospective companies that I identified and encouraged the officers to do likewise. I received help from the senior officers — chairman, president, chancellor, provost, vice presidents, etc. Within a year, we had turned the tide. The gross income, as I recall it, went up from $860,000 per year to $1,240,000 and by the end of the next year, we broke $2,000,000.

The excitement became contagious. I had made an agreement, after we had stemmed the tide of member company losses, with Chancellor Gray, whereby we would add one new officer plus one/half secretarial slot for each fifteen new companies. Space would soon become a problem, but one that could be resolved. I was aided greatly by Ms. Janet K. Anderson, who came over with me from the department.

Meanwhile, I had cut down on my departmental activities. I retained the chair plus the responsibilities for the U-P Lectureship, together with the other members of the committee. However, my days of increasing research volume and an increased number of graduate students ceased. I felt that I could do more for the Institute, that I loved, in this way. The Institute needed not only increased sources of revenue, but also increasing interaction with industry, which would lead to increased industrially sponsored, research volume, increased fellowships and scholarships, and more visibility.

As the numbers of companies from American industry grew, so they grew in Europe. My trips during the 1970's, and 1980's began to pay off in terms of increasing interaction with European industry. It was not only my efforts alone, but also, and especially in the early 1970's on, with help from particularly Eric Johnson and Laura Scott-Stout.

We began to run symposia and special "summer" one and two week courses in Zurich and elsewhere. The program kept on growing as a result of the combined efforts of faculty, administration and staff members of the program. As Director of Industrial Liaison, I had responsibility for interactions with industrial concerns in addition to the I.L.P.. But, I concentrated my limited time and resources in the I.L.P. and what flowed therefrom.

Beginnings of the I.L.P. Program in Japan

When I became Director of Industrial Liaison, Yaichi Ayukawa, my former student and close friend for two decades remarked that we should get moving in Japan. To do so successfully would require an M.I.T. presence in Japan. He suggested a small office in Tokyo attached to his TechnoVenture office. General Lampert was in complete agreement and slowly and carefully brought his fellow senior officers — particularly, the Chairman, President, and Chancellor aboard.

At an M.I.T. Corporation meeting in 1976, authorization was granted to General Lampert to have an M.I.T. office opened in Tokyo. It was located in the Diamond Plaza Building as part of Ayukawa's suite. A few years later, because both we and TechnoVenture needed more space, we moved to the present location nearby. Ayukawa continued to be very helpful. While the Corporation deliberations about the M.I.T. office were going on, Yaichi and I signed up four new member companies from Japan. This would be enough to cover expenses for the first year and we could be assured of keeping our promise to break even, at the very least. But we grew very rapidly.

Our first Japan office director did not prove satisfactory and Jim Lampert, in 1978, introduced me to George Sankey, a retired Colonel who had served as his language aide in Okinawa. He was a find as he was capable, articulate, bright, and knowledgeable in Japanese customs, as well as having an extreme fluency in the language. One could write a book about George. He passed away less than a year ago and I still miss him very much. His wife Anita, Diana, he and I became very friendly over the years. George and Anita did much for M.I.T. during his tenure as Director of the Tokyo Office, as well as for the decade he worked as Special Assistant to the Vice President after they moved to the Los Angeles area, where they had purchased a home in North Hollywood. He continued to travel to Japan whenever I did or with other senior officers of M.I.T., and made many friends there, not only his own, but for M.I.T. as well.

In 1977, we held our first major symposium in Japan at the Keidanren. Jim Lampert chaired it and the speakers included Paul Gray, Walter Rosenblith, Charlie Kindleberger, Bill Pounds and myself. It was a huge success. Some of Jim Lampert's friends in the Japanese Diet came and spoke — invited or not. It all boded well for us in Japan.

We hired a number two person in the office, Mrs. Akemi Saito, who is there to this day and does a magnificent job. Jim Lampert traveled back and forth to Japan twice per year as did I, and together we built up the membership all under the guiding hand of, tutelage of, and assistance from Yaichi Ayukawa. We owe much to him and to George Sankey — alas both since departed.

Jim Lampert, in the Spring of 1978, developed lung cancer and was gone within a matter of months in spite of the best treatment at Walter Reed Hospital in Washington. He was a great man, a wonderful individual who was a "courtly man" as Howard Johnson stated at his eulogy. Jim fought the disease to the end. I recall his calling me at the Melrose Wakefield Hospital, where I was recovering from internal bleeding due to too much coumadin and not having my prethrombin time checked, because I was on a trip to Minneapolis. He cheered me up that June and he was pretty well gone himself. He passed away in July, 1978.

Jim's passing left a void in Resource Development. We were in the midst of a $225 million Capital Campaign. Chairman Johnson, and Drs. Wiesner and Gray had their work cut out for them and I was busy with the I.L.P. It was still growing, at home, in Europe, as well as in Japan.

Family Affairs

Shortly after moving to Melrose and after joining the William Underwood Co., Judy's husband began evening studies at Boston University leading toward an M.B.A. degree. He persisted and worked hard by day and by night and obtained his M.B.A. Degree from Boston University in May, 1977. Now he had the proper academic tools and credentials for his chosen career in business. We were very proud that he had the enduring power to see this through to the end, inasmuch as I knew how difficult it was to teach courses in the evening after working all day, and how much more work it was to take the courses for credit with family obligations, etc., which Rich never shirked. Our hat was off (and still is) to him for this achievement.

On June 1, 1975, Jonathan graduated with his class at Emerson College. We were delighted. He went to New York and obtained a position as night clerk at the Chemists' Club of which I happened to then be a member. (He got the job entirely on his own initiative). I learned about it after the fact. It gave him a chance to try out for shows, plays, etc., in the "Big Apple."

Meanwhile, I should state that Jonny had developed a superb tenor voice and was singing in the Temple choir and had acted as Cantor at Lisa Tannenbaum's Bat Mitzvah. He was splendid! More about his singing later on herein.

After I recovered from the internal bleeding mentioned above, Diana and I had planned a trip to Europe. I was fully recovered and she insisted that I would gain back weight and strength better abroad than at home, where I would be forever on the telephone. She was right. I had to be in Europe on Nestle business, as well as to make some calls for the I.L.P.

Nestle had given me some confidential assignments and I was then consulting for them exclusively. One of the assignments was to visit a laboratory they had in Normandy. We flew to Paris and took the train to the Normandy region. It gave us a chance to tour Normandy, the beaches, Avaranches, St. Lo, etc., and Diana saw the magnificent tapestry at Bayeux. It was a memorable trip.

When we returned from Europe, I was like a new person. I had regained the lost weight and strength. It turns out that I would need it.

Miscellaneous

During this period of 1974-1978, I developed a close friendship with the Lord Zuckerman. Solly had served as in Whitehall during and since World War II and until his death. He had been the Science Adviser to Air Marshall Tedder during the war and then served the various governments of England as Science Adviser, since around 1954 until his retirement. I knew him well during the 1970's and until his death early in the 1990's. Diana, he and I had many a meal together. I knew him well through his being on the Board of Cadbury Schweppes and his introduction of Cadbury Schweppes to me, and my meeting Sir Adrian and Domenic Cadbury through him. Ergo they became members of the I.L.P. Solly preceded Greg Henderson as the President of the Royal Zoological Society in Regent's Park.

His lasting (for a while) legacy to me was his introduction of me to his personal tailor, Mr. L. Minc, on Dover Street, London W1. I still have a suit and a jacket that Mr. Minc made for me. They are beautifully tailored. The problem was that I just put on too much weight. Mr. Minc's clothes fit well and lasted!

Through Bob Hagopian, Director then of Corporate Relations at M.I.T., I met Mr. Hugh Parker, the Founding Director and then in charge of the London Office of McKinsey and Co. Hugh was very helpful to us and to me in the I.L.P. and through him, we picked up many new member companies for the I.L.P. in the United Kingdom.

Hugh came to see Diana and me off in June 1982 when we took the Orient Express from London to Milan, bringing us, in addition to his warm friendship, a bottle of champagne for the trip.

I have remained close friends with Hugh and especially since he returned to live in Marblehead, Mass. We see each other or talk to each other at least once per week.

If one were to ask me how we managed to build up the program to the extent that we did in that period of time, I would say that there were several key elements. These were: the cooperation of the senior officers who, in the course of the Leadership Campaign travels and commitments, never forgot to sell the I.L.P. The second was the cooperation of the faculty as mentioned above. The third was the help from alumni such as Hugh Parker in England and Dr. Yaichi Ayukawa in Japan. The fourth was the internal excellence of work in servicing the member companies by all of the I.L.P. staff and Bob Hagopian, then Director of Corporate Relations, in bringing in new member companies, the faculty, and the I.L.O.'s taking the initiative.

Through Arnold Spicer's introduction, I became friendly with Sir Ernest and Lady Chain. We spent a number of Friday nights at their home and he and I did some joint consulting in the 1970's for Nestle. He participated in the Freeze Dehydration and Advanced Food Technology Symposium arranged by Louis Rey, Walter Rothmayr and myself in the Burgenstock, overlooking Lake Lucerne in 1973. The Proceedings of the Meeting were published by the Academic Press in London in 1975 and were well received.

Diana and I attended a Symposium in June 1976 in honor of Sir Ernest's 70th birthday. It was a three-day affair with formal dinners at the Middle Temple, dinner at the Chain's home in Wimbledon for the overseas guests, and a dinner at the London Hilton for the Technion on the last night. I recall the immense security at the symposium which was held at the Royal Society. This was due to the presence of Prof. Ephraim Katchalski-Katzir, who was then the President of Israel and who was one of the Symposium speakers.

In 1970, I was elected one of the first group of Fellows of the Institute of Food Technologists (FIFT) and in 1974 was elected a Fellow of the (British) Institute of Food Science and Technology

(FIFST). I believe that I have elsewhere covered my election as a Foreign Member to IVA, the Royal Swedish Academy of Engineering Sciences, and Foreign Membership in the Swiss Academy of Engineering Sciences on October 17, 1986.

CHAPTER 17

VICE PRESIDENT FOR RESOURCE DEVELOPMENT (1978 - 1986)

General Lampert unfortunately passed away on July 9th, 1978 after a long battle against cancer. We attended his funeral at West Point, on July 13th, 1978 driving there with the Stuart Cowens. It was as one can imagine, a most impressive ceremony. One of his sons conducted the service.

All during the remainder of July and throughout most of August, the rumor mill was in action in Resource Development at M.I.T. as to who would succeed General Lampert. There were two inside obvious (to the staff) selections and then the unknown from the outside of the Institute. It was felt that nothing could happen for awhile, inasmuch as the three principals who would be involved in the decision of the succession were not around the Institute all summer — at the same time.

In the event, Dr. Gray summoned me to his office on the morning of August 31, 1978 and asked me to be Vice President for Resource Development. For the next few days, he asked me to inform no one other than Diana, of course, whom I asked to speak with before my final agreement. This I did and we shook hands on it. We kept mum. On September 1st-4th, we spent the Labor Day weekend with the kids at our camp in Northwood, New Hampshire, closing it up for the season. Diana and I kept quiet about the momentous decision we had made. By September 7th, the Academic Council (the senior administrative body of the Institute made up of the President, Provost, Chancellor, all the Deans and all the Vice Presidents) had been informed. I was now free to talk with anyone about the appointment. On the 8th of September, Howard Johnson, our Chairman of the M.I.T. Corporation, was kind enough to call me and inform me that the Executive Committee approved my new appointment enthusiastically.

On the afternoon of September 8th, I attended a beautiful Memorial Service for Jim Lampert at the Kresge Chapel. On September 12th, 1978, I attended my first meeting of the Academic Council.

First Duties as Vice President

My first assignments as vice president were to make certain the current $225 million Leadership Campaign remained on target and to familiarize myself in detail with the Campaign staff and to speak with all of them about plans, based on notes that I had jotted down over the long Labor Day weekend as I thought about the matter. I had to reassure the staff and yet make certain that there was no complacency. We had to move forward. In this matter of the Leadership Campaign, one could not have asked for better cooperation than that I received from Nelson (Ned) C. Lees. Nor, could one have asked for more or better help and leadership than I received from Howard Johnson, Jerome Wiesner, (who would soon spend more than half of his time on campaign business), and Paul Gray. I felt that the second and immediate problem to be tackled was that of my successor as Director of the Industrial Liaison Program. Paul Gray identified Prof. James Bruce, then Acting Dean of Engineering (later to be appointed Vice President for Information Systems). Jim took over as Director of the I.L.P. and did an excellent job. When he became Vice President of Information Systems, I appointed Prof. James M. Utterback as Director of the I.L.P.

Now I had the systems in place and with the hard work and plentiful help of key volunteers, the above mentioned senior officers, and the hardworking staff, the Campaign exceeded its goal.

For Diana and me, it meant many trips around the United States and to Japan and Europe, and these trips continued unabated even after the campaign ended in 1980, because campaigns are expected to (and do, if nurtured properly), increase the annual level of eleemosynary gifts to an institution.

FIG. 17.1. YAICHI AYUKAWA AND SAG AFTER RECEIVING BRONZE BEAVER AWARD FROM M.I.T. ALUMNI ASSOCIATION, 1980

Other duties as Vice President of Resource Development included attendance at all meetings of the M.I.T. Corporation, serving as an exofficio member of its Development Committee, membership in the Committee on Resources of the Institute and of the Academic Council, which then met every Tuesday morning — usually through lunch. These latter were interesting meetings dealing with policies of substance of the Institute, promotion, tenure, etc. It was chaired by the President or in his absence by the Provost. During the remainder of the Wiesner Administration, Walter Rosenblith continued as the Provost. During the first five years of the Gray Administration, Francis Low, a theoretical nuclear physicist, was the Provost and he was followed by Prof. John M. Deutch, who is now the Director of the C.I.A. in Washington. All three were different. Each one had his strong points. Rosenblith was a lovable person (for me) and bright, articulate, dogmatic in many respects, but a first rate scholar and good judge of people 99% of the time. But, when he backed the wrong horse, that person was lucky and many others suffered. He and Jerry Wiesner, whom I also loved and respected, and had known almost since his arrival at the Institute, were an excellent pair and were well-suited to and complemented each other.

Francis Low was an entirely different sort of person. Bright, but low key, outwardly appearing to be easy to manipulate because of his even temperament but, in fact, solid as a rock in his beliefs. To me he was the most ideal person as Provost. Then came John Deutch, brilliant also, articulate, never fearing to open up, who gave the impression that he shot from the hip. I got along well with all three but my favorite was Francis Low. However, I cared for and respected them all and got along with them all, as well.

Francis brought in Dr. Kenneth A. Smith, a Professor of Chemical Engineering, as Associate Provost and Vice President for Research. I feel that he was ideal for the job and we became good friends to this day.

Early in the Gray Administration, President Gray and I went to Japan on a business trip, to attend a meeting of the M.I.T. Club of Japan, and many other such affairs including calling on many companies, holding a reception, etc. It was quite a trip and we capped it off by the return trip to M.I.T. where, (in those days, one went via Alaska for a one-hour stop to refuel, etc.), we met Lawrence Rockefeller whom I saw on the plane. He is a member of the M.I.T. Corporation (now emeritus) and we three had a good chat on the plane and later at Alaska. He had been to Japan with some of his people to raise money for Sloan Kettering. As a result of this trip he started coming to some M.I.T. Corporation meetings again and became interested in our philosophy program at M.I.T. He soon donated a Professorial Chair for the program which was filled ably by Thomas Kuhn.

The visibility of the Vice Presidency, I am sure, assisted in my being asked to be on a number of additional Boards of Directors. I accepted only two of note. One I have mentioned elsewhere was the High Voltage Engineering Corp. The second was

FIG. 17.2. RETIRED R. ADM. YOSHIO KUBOTA (M.I.T. '23) AND SAG AT M.I.T. CLUB OF JAPAN MEETING IN EARLY TO MID-1980'S

later in about 1982 at the Ionics Co., in which I am still a Board member at the time of this writing, mid-1996. I was a member of the Board of Directors of the Florasynth Co., since the 1970's. I have known Lorraine and Jack Friedman since 1958 and we were the recipients of many kindnesses from them — staying at their Florida apartment adjacent to the Breakers in Palm Beach many times in the 1970's and 1980's, as well as many other kindnesses to me and to M.I.T., and I remained on the board those many years. The company was sold in 1993-1994. Lorraine and Jack deserved some rest. They have built up the company Lorraine's father founded and I am fortunate to have such warm friends.

During that period as vice president of M.I.T., I went to Japan, as a rule, twice per year and during that time frame (1978-1986), and even later, George Sankey was usually with me. I will relate herein mainly personal activities that might be of interest. I believe that I have elsewhere herein related the story of the Symposium and Dinner on my behalf after I became Vice

President. This took place on September 24, 1979 and that weekend, Diana had a big party at our house for our overseas guests who came and which was quite an affair.

FIG. 17.3. DIANA, RICHARD CLARK (SON-IN-LAW) AND SAG AT SYMPOSIUM AT M.I.T. ON SEPTEMBER 24, 1979 IN HONOR OF SAG'S 60TH BIRTHDAY

Revisit to Roku Roshi POW Camp, Fukui-Ken (Prefecture)

In late 1979, I mentioned to George Sankey, who was still living in Japan with Anita, that I had a "yen" to visit just one more time two former P.O.W. Camps, if they were still extant. George in his usual thorough manner, arranged for the first of these to take place in March 1980 by our taking a side trip from Kyoto to Roku Roshi. Diana was with me on that trip and Anita came along also. We had a great time. We left Kyoto on a Saturday in mid-March, 1980 by train alongside Lake Biwa to Fukui. There, we were met

FIG. 17.4. ABOARD TRAIN ENROUTE TO FUKUI TO VISIT ROKU ROSHI POW CAMP SITE: COL. GEORGE K. SANKEY AND DIANA, MID-MARCH 1980

by a newspaper reporter for whom George arranged. We were taken by car to Ono where we were met by the mayor of Ono. After some introductory comments, I presented the mayor with a modern small transistor radio to replace the one purloined by one of our men near the end of the war. The mayor of Ono presented me with two small dolls made in the area. They are emblematical not only of the local crafts but also of the head covering used against snow and rain in the mountains.

We then journeyed by car up the hilly road to the village of Roku Roshi. While at the Ono City Hall, we had met the only person who was employed at the camp, a then 84-year-old gentleman who was a guard — as I recall it, a "friendly" guard. He, whose name I do not recall, went with us to Roku Roshi. The only visible remnant of the camp is the old kitchen building, then used

FIG. 17.5. AT ONO CITY HALL, RETURNING A SMALL RADIO TO MAYOR OF ONO, FUKUI-KEN, JAPAN

FIG. 17.6. FORMER GUARD AT ROKU ROSHI

Note snow on ground in Roku Roshi village.

as a storehouse. In the distance, we could see the gravesite of the one man we lost at Roku Roshi, James J. Mollette. The whole area, other than the roads, was covered with eighteen inches to two or three feet of snow. I'm sure we would never have lasted through a winter up there without heat. I understand that the area is now a popular ski resort. Following photographic coverage, as George Sankey used to say, we left to return to Kyoto via Ono and Fukui. It was a nice but nostalgic weekend.

FIG. 17.7. DIANA AND I AT ROKU ROSHI WITH COOKHOUSE BEHIND US, MID-MARCH, 1980

FIG. 17.8. ANITA SANKEY, GEORGE SANKEY, DIANA AND SAG IN KYOTO AT END OF DAY SPENT ON ROKU ROSHI VISIT

A Lawyer in the Family

In 1977, our daughter, Judy Clark began Boston University Law School. One important aspect was to have someone at home, theirs or ours, when Rachel got through kindergarten at noon during school days. Diana volunteered. She and Rachel hit it off and Rachel never had as big a fan as Grandma Diana. Diana took her with her to visit Grandma Ida and Great Uncle Max every Tuesday afternoon, as well as to the North Shore Music Theater on Children's afternoon, as she had done with Sarah several years earlier. Diana and I both admired the perseverance Judy and Rich both had through the three years, 1977-1980 and the cooperation Judy received, especially from Rich. It was great. Judy graduated with the Degree of Juris Doctor in June, 1980 and proceeded to study for the Bar Examinations which she took in the early fall.

Diana and I were at the Excelsior Hotel in Florence, Italy when we received the joyful news that Judy had passed the Bar Exams. In due course, in a month or so, Diana and Rich had the pleasure of witnessing Judy being sworn in as a member of the Massachusetts Bar.

Judy had been clerking for Thomas Sullivan, former Mayor of Melrose, after classes in her senior year and thereby hit it off so well that she was offered a position at Sullivan and Limone upon graduation and later became a partner of an expanded law firm. Tom Sullivan, a great man and a superb lawyer, died a few months ago from cancer as a result of asbestosis he incurred in the Navy in the Korean War, and Judy is the senior partner of Sullivan and Clark, and is highly respected in Melrose, as is her husband Rich who is the Comptroller of the Xenergy Corp.

FIG. 17.9. JUDY'S GRADUATION FROM B.U. LAW SCHOOL, DIANA, SARAH AND JUDY, JUNE, 1980

FIG. 17.10. JUDY AND DIANA AT B.U. LAW SCHOOL GRADUATION
Jonathan in right background.

Revisit to Zentsuji

George Sankey made arrangements for the two of us to visit Zentsuji on Saturday, November 14, 1981. This was quite a "production." He had made the arrangements with a friend of his at NHK, the government broadcasting station, (the PBS of Japan). On the morning of Nov. 14, we flew from Osaka, where we had been visiting various companies and individuals, to Takamatsu on Shikoku Island. There we were met by the reporter and cameraman (movie) of NHK and crew. We drove to Zentsuji, and more coffee stops. I found out later that the coffee breaks were to use up time, because they had arranged for Capt. Yuhei Hosotani (former Commandant of Zentsuji POW Camp) to meet us at a certain hour later in the day. At any rate, I was anxious to see the ex-camp.

They drove us to the general area and pointed out a certain building which I stated was not the place. I asked George to hold them at bay while I reconnoitered. I found my landmark, an ancient Shinto Temple, five-tiered. I called them over and showed them where the POW camp had been on the other side of an alley or lane that ran alongside the temple grounds. There was a new junior high school there. I said "let's go in and check." Sure enough, a few years earlier, this new school had been built on the old site after the buildings had been demolished. I had a wonderful talk with some of the students and teachers and stated that this was a far better use of the land than the earlier use as army barracks and POW camps. Meanwhile, the NHK cameraman was filming away.

At 2 PM on that Saturday, we were driven to a certain spot a mile or two away and there was Captain Hosotani. We shook hands and talked a bit with George translating. Hosotani also agreed that the Junior High School was now where the camp had been. Then we repaired to the local town hotel/restaurant where NHK had prepared a private room and there was a large group of former guards, etc., all arranged by Capt. Hosotani's former secretary who was there. She had been a part-time secretary and worked for the POW Camp after school hours. It was quite a spread but I found it hard to eat much. There were three former civilian employees of Hosotani at the meal including the former medical corpsman, and the bookkeeper/accountant.

Later after dinner, George and I were driven to Takamatsu where he had arranged for a hotel and we walked through a shopping arcade that was still open. Early Sunday morning we took off by plane for Osaka and back to work. Then on Tuesday, I went back to Tokyo with George and returned home on November 18th. It was a long trip. It too brought back memories of forty years earlier. Not too long after I returned home, I was sent a video of my trip by Mrs. Saito to whom it was directed by NHK for transfer to me.

The video made by NHK was shown on their network. It began with Japanese fighter planes bombing Bataan, the Death

March, and then showed us at dinner in Zentsuji. Further on in the video, Captain Hosotani was shown at his home declaring that he had done no harm as a soldier and was only doing his duty. He claimed that he had not done anything to warrant his being sent to Sugamo Prison for five years as a War Criminal. Well that may be, but I could cite from memory a few things that were not so pleasant. However, George and I were set up for this, unwittingly. Apparently, it was Captain Hosotani's price for appearing with us. But it was good to get back again to Zentsuji to talk with the students and teachers. It was the best POW camp of the lot that I had been in. Captain Hosotani was on this remeeting a teacher of English at a Junior High School in Takamatsu. He apparently was fluent in written English but not in verbal. He had no one with whom to converse, apparently.

Visit with Ayukawas to Australia and The Great Barrier Reef

Yaichi and Masa had been after us for some time to join them on a trip to Australia and more specifically to The Great Barrier Reef. He wanted to reward me in a non-monetary fashion for what I had done for his company, TechnoVenture, for which I had taken no fees as a director nor expected any. The Ayukawas were dear friends. But they insisted and we went with them to Australia, where neither Diana nor I had been before.

In the event, we joined Masa, Yaichi and Junta to fly to Australia when my M.I.T. work was done on a trip in early March, 1982. On March 18th we flew all night from Tokyo to Brisbane. We stayed for one night at Brisbane where we saw the Kuala Park and Kangaroo Park and Yaichi did a small amount of business. Then the next day, we flew in an old Sikorsky helicopter from Cairns to Lindeman Island where we stayed for two days, I believe, and then on to Haymann Island in a small plane. It was beautiful scenery. On the 22nd we flew to Sydney, where I looked up a friend of our ophthalmologist, Dr. Alpert, a Mr. Berman who was the Australian representative for the Norton Company of Worcester. His wife and he introduced Diana and me to a lovely club for cocktails and dinner. It was a grand evening and they showed us all around Sydney et environs.

The next day, we took Junta Ayukawa on a boat trip of the Sydney Harbor. We loved the large suspension, single-span bridge across the harbor which was fifty years old that year and we visited the beautiful new Sydney Opera House. That afternoon, Diana and I went to the Great Synagogue of Sydney of which the Minister was Rabbi Raymond Apple. The "Rebbetzin" showed us and another couple around. It was a truly Orthodox Synagogue, "bima" in the middle. It was a delightful day. On the 27th, we flew back to Tokyo, overnight and thence home on Saturday.

Trip Aboard the Orient Express, June, 1981

In June, during a regular trip to Europe, I decided to give Diana (and me) a real treat. As much as she hated flying, so she loved train travel. I had booked us some months earlier on the Orient Express from London to Milano via Paris and Lucerne. We had arrived in London on the morning of June 12th, and transferred to Paris that day, and I visited some laboratories and saw Nestle scientists/engineers in Paris until the 16th, when we went back to London, where I had some business with Ranks, Hovis and MacDougall and at Unilever on the 17th and 18th.

On Saturday, June 19th, we spent the day with the Hendersons at Reading and on Sunday, June 20th, repaired by taxi to Victoria Station to board the Orient Express. Hugh Parker came to see us off laden with a bottle of champagne. At 11:44 AM, the British section of the Orient Express train departed Victoria for the Channel Port. The British Section consisted of all dining cars, as I recall and we ate all the way to the Channel. It was a most sumptuous meal, by far the best that I have ever been served on any railway journey — excellent food, elegant service, white linens, etc. We sat in Edwardian era or early George V era vintage cars, with polished wood paneling. It was gorgeous.

We walked from the train to a ferry for the relatively short trip across the Channel to Boulogne where we then boarded the

Continental Section of the Orient Express bound for Paris, Lucerne, and Milano with a final destination of Venice, I believe. We got off the train the next morning at Milano.

The evening meal was also sumptuous with many people dressed in dinner jackets and long dresses (we were not). We had a bedroom, consisting of a sitting room by day. At night, it had an upper berth that pulled down as well as a lower with a sink opposite the sleeping facilities. If one needed to use the lavatory, there were facilities at one end of the car. At the other end, the porter had his cubbyhole where he made coffee and served croissants and coffee for breakfast the next day. The croissants were picked up at Lausanne at 6:30-7AM. It was quite an experience in the old Wagon de Lits. I know that Diana loved it as did I. We stayed overnight in Milano at the famous Hotel Excelsior and Gallia and on the morning of the 22nd headed back for Lausanne on the picture window train. It afforded a gorgeous view of the scenery that we had missed the night before as we came through the pass. The Alps, in sunlight, are simply beautiful in June. Then it was back to appointments which Nestle had arranged for me until July 1st.

On Sunday June 27, we met George and Anne Buchi at Freddy Giaradet's famous Michelin two star restaurant, a memorable event! On July 2nd, we left Zurich for home.

During the period encompassing the late 1970's and early 1980's, I served as Chairman, part-time of Fisheries Research & Development Ltd., (arranged for by my good friend, Mr. Paul Jacobs). This involved my being in Halifax for four meetings per year. The Director and President of the company was a former student (who is still there), Dr. John F. Maloney, Jr. It was quite an experience going to Halifax and Lunenberg and Diana joined me on at least two or three occasions. The weather at Halifax can be a bit "dicey" on occasions. One can see how the "rum runners" slipped in and out of the fog banks during prohibition.

The years of my M.I.T. Vice Presidency were busy ones — meetings every day, visitors and a great deal of travel to prospects.

For the main part, I have selected some of the fun times for this memoir. It was not all foreign pleasure travel by any means. I have just highlighted some of these because they afforded Diana and me great pleasure.

In 1982-1983, Diana had undertaken a large needlepoint project for Yaichi and Masa Ayukawa. She wanted to give something to them that they could not purchase. She made a large needlepoint or perhaps crewel. It was a work of two ducks and we had it framed and "shlepped" it to Japan for our April visit. They loved it and had a silver plaque placed on the frame. We then went to the Hakone Prince Hotel with the Ayukawas for a weekend. It was gorgeous scenery and lavish. The weather was beautiful and we could see the top of Mount Fuji set between two smaller mountain peaks. The hotel was above Lake Ashii, a beautiful

FIG. 17.11. DIANA PRESENTING YAICHI AYUKAWA A SPECIAL PIECE OF HANDWORK SHE HAD MADE FOR HIM (APRIL 1983)

spot. It was from here that I called my dear friend, Dr. Michael Kane, at M.I.T., when I was having some trouble with my hypertension medication.

Death of Ida Frances Greenberg, Diana's Mother, June 14, 1983

On Flag Day, June 14, 1983, Diana's mother passed away at Willow Manor Nursing and Retirement Home at about 8:30 PM. She was taken to the Lowell General Hospital where Diana saw her for the last time that evening, but she was already gone. A wonderful person and a great mother, wife and grandmother. She was and still is sorely missed by her descendants. The only one left of her generation then was Uncle Max B. Greenberg, George's youngest brother.

FIG. 17.12. SARAH BETH CLARK AND GRANDPA SAM
Grandpa needs to practice more.

1984: A Year Full of Simchas

Following my trip in March, 1984 to Japan, Diana and I proceeded to Singapore, where neither of us had been before, to visit Eastreco for the Nestle Co., and coincidentally to attend the opening of this new laboratory. It was *very* hot in Singapore and we did not get around too much. We did observe the new and gorgeous airport at Singapore and noted the long, long flight home from Singapore to Hawaii to San Francisco.

Sarah Beth Clark, our elder granddaughter, celebrated her Bat Mitzvah on May 4, 1984, Parsha Emor. She did a magnificent job as did Judy and Rich in matching her performance with the catered Oneg following thereafter.

FIG. 17.13. EMMA AND ROBERT CLARK (RICH'S PARENTS) WITH "AUNTIE" GAIL GOLDWASSER AT SARAH BETH CLARK'S BAT MITZVAH, MAY 4, 1984

FIG. 17.14. RICH CLARK (THE HAPPY DADDY) AT THE ONEG FOLLOWING SARAH'S BAT MITZVAH

FIG. 17.15. TOM SULLIVAN, JUDY'S LAW MENTOR, AND JUDY CLARK AT THE ONEG, MAY 4, 1984

FIG. 17.16. RABBI STUART POLLACK OF TEMPLE TIFERETH ISRAEL AND RABBI HERMAN POLLACK AT ONEG FOLLOWING SARAH'S BAT MITZVAH

FIG. 17.17. MR. AND MRS. JOHN TAMMARO, LONGTIME FRIENDS AT ONEG

The next day, May 5, 1984, was my 65th birthday and Diana ran a big catered party at the house in celebration. It will long remain in my memory as the most magnificent birthday party ever produced for me.

On May 19, 1984 our son, Jonathan, was awarded the Master of Science Degree in Management at Lesley College where he had been attending late afternoon and evening classes while employed at Polaroid. We were so proud of him inasmuch as this was entirely on his own initiative and we knew how, at an earlier time, he disliked school, etc. Through my dear friend, George Buchi, I obtained a Rolex (wholesale) in Switzerland for him, a few months before graduation. He wears it with pride to this day. A superb product for a super son!

FIG. 17.18. SAG THANKING DIANA FOR A MEMORABLE 65TH BIRTHDAY PARTY, MAY 5, 1984

FIG. 17.19. TWO LONGTIME FRIENDS, GEORGE SEYBOLT (L) AND MILTON ISSENBERG (R) AT 65TH

FIG. 17.20. GOOD FRIENDS AND COLLEAGUES, GEORGE BUCHI (L) AND BRIAN SUTER (R) OF NESTLE AT 65TH

FIG. 17.21. JONATHAN AND DIANA AFTER HIS RECEIVING HIS MASTER'S DEGREE FROM LESLEY COLLEGE, MAY 19, 1984

On Friday, June 1st, 1984, in the presence of my family and friends, I received the Second Class of the Order of the Sacred Treasure from the hands of the Consul General of Japan in Boston, H.E. Sadakazu Taniguchi. The Ayukawas, both Masako and Yaichi, came for the ceremony. As a matter of fact, I feel that it was he who made sure Mr. Taniguchi's predecessor, Ambassador Takeo Iguchi, nominated me for the high honor. The award had been announced on the Emperor's birthday, April 19, 1984, in recognition for my efforts to build bridges of mutual understanding and friendly relations between our two countries through student exchange, my lecturing in Japan, etc. I had a large number of students and post docs from Japan over the years and it seemed to be a good and natural thing to do to build these bridges based on knowledge and love rather than reiterating all the past wrongs which could only lead to more bloodshed.

FIG. 17.22. SAG BEING AWARDED THE ORDER OF THE SACRED TREASURE
L to R, Diana, SAG, Consul General S. Taniguchi and Mrs. Taniguchi, June 1, 1984.

Later that evening, Consul General and Mrs. Sadakazu Taniguchi had a special dinner at their home for us and our children, Judy, Rich and Jonny as well as the Grays and the Ayukawas. It was quite a meal and the day was the beginning of a friendship with Tani and Yuki that lasts until this day. He was the Ambassador to Israel when Diana passed away and thoughtfully he sent me some stones from the Mount of Olives as well as a container of dirt. What thoughtfulness! The letter he wrote was characteristic of their warmth and friendship for Diana and me. He later served in Osaka as the Ambassadorial representative of the Gaimusho (Foreign Ministry) and most recently served as Ambassador to New Zealand. I believe that he is now retired from the Gaimusho. It is hard to believe that he is old enough to retire. After he had been in Boston and returned to Tokyo, we visited with them when we could. But like other government jobs all over the world, a posting lasts three or at the most four years.

FIG. 17.23. THE GRAYS AND THE GOLDBLITHS AT THE INVESTITURE OF THE ORDER OF THE SACRED TREASURE

Several Trips to Europe in Summer and Fall, 1984

When Diana and I were in London in late June and early July 1984, Ken Durham, a friend of long standing had heard that Diana liked to watch tennis matches, and in fact she would spend most of the afternoons in the hotel room while in London watching the matches at Wimbledon, as she ate sandwiches she had picked up in the AM while shopping at Marks and Spencer. (She also brought me back some beautiful strawberries and heavy cream, Yum!). Ken Durham offered us the use of his chauffeured Rolls Royce and two choice seats at Wimbledon for Friday, 29 June, 1984 (this was repeated in a later year while he was still chairman). Diana was in seventh heaven. I enjoyed the experience also, although I must confess that the private area for smoked salmon sandwiches and strawberries and cream interested me more. It was quite a day.

On July 2nd, George Buchi, Gerry Wogan and I were taken to Unilever to make a research presentation in order to try to raise funding for toxicological research in the department. We succeeded but they worked hard!

On Tuesday, July 17th, I alone went back to Europe again for the Nestle group and traveled with one of their attorneys, Mr. Ferdinand Elsener, from Zurich to Stuttgart for Ludwigsburg, to Munich, (where I sent back the butter at the Hotel Vier Jahreszeiten because it was rancid), to Milano (Excelsior Hotel again), Paris, Beauvais (France), Lisbon, Zurich, Vevey, and finally home from Zurich on July 28th, 1984. It was quite a busy trip.

On August 8, 1984, my good friends, Messrs. George Buchi, George Seybolt, Dick Mateles, Marc Karel, Dan Wang, Chris Walsh, Steve Tannenbaum, Sandy Miller, and Gerry Wogan threw a black tie gourmet party at Mansion Robert in celebration of my 65th. It was quite an event in a busy summer, inasmuch as I had many duties as V.P., other than traipsing around Europe and attending parties. It was a memorable dinner — hand expressed menus, etc. They presented me with a set of the Encyclopedia Judaica, which I treasure and use!

I had to go to Europe one more time for a week in October on this special mission for the Nestle Co., mainly in Switzerland this time, and then one more trip to Japan in November for two weeks with the Grays, and that about concluded my travels for the year. It had been a busy one! In late December, Diana and I took a week off at Christmas time and went to Bermuda which we both loved. We met our good friends, Marvin Goldstein and his wife, Connie Lewis, whom we have known since we moved to Melrose. Marv unfortunately died not too long afterwards from cancer. He was a bright and very nice person whom we enjoyed being with.

FIG. 17.24. TWO M.I.T. PRESIDENTS IN JAPAN, JEROME B. WIESNER AND PAUL E. GRAY, CIRCA 1984

Some of My Activities in 1985

In 1985, the hectic pace continued with several meetings each day, or trips within the U.S.A. or abroad to Japan and Europe. One such trip bears mentioning and that was of Governor Michael Dukakis who was going to China and Japan to tell the respective governments and industrial concerns, especially in Japan, about the reasons they should build plants in Massachusetts and, in general, to invest in Massachusetts. This took place in March 1985, when I was to be in Japan on behalf of M.I.T. anyway. I was invited as part of the Dukakis mission to visit China as well as Japan. I declined on the China portion, but assisted him on opening doors for him at various industrial concerns in Japan. I also set up a reception for him at the Hotel Okura and invited the top people in the major Japanese industrial concerns, as well as influential members of the National Diet to meet the Governor of Massachusetts. Most came, as did Ambassador Mansfield, who spoke also. The Toast ("Kampei") was given by the former Japanese Ambassador to the U.S. and USSR, as well as former Justice of the Japanese Supreme Court and Commissioner of Baseball in Japan, the Hon. Takezo Shimoda, another good friend in Japan.

Michael Dukakis was so appreciative of the help I (and George Sankey) and Fred Gross gave him that he attended the next meeting of the M.I.T. Corporation to publicly express his appreciation for our services to him and the Commonwealth. (Dukakis had always been invited to the meetings of the M.I.T. Corporation, inasmuch as the Governor is an ex officio member, but this was the first time he had attended). One of the key persons in making the Dukakis visit to Japan so successful, in addition to George Sankey, was Mr. Fred Gross, Senior Industrial Liaison Officer, who was responsible for Japan in our I.L.P. office. Fred had worked at the State House for part of his career and knew Governor and Mrs. Dukakis very well and he was (and still is) a most dynamic individual. He was a tremendous asset and was key in making the trip a success for Dukakis and the Commonwealth. Diana and I were and I am very fond of Fred and Laurie and their two children. We spent a great deal of time with Fred in Japan and I hear from him frequently from California where he is Executive Director for Development of the Rand Corporation. Fred and I broke bread together many times during the last half of the 1980's. These afford pleasurable memories.

On April 12, John Toland, the Pulitzer Prize author and historian, was to be in Boston on a book signing/publicity tour. I had always admired his writings and most especially, his then recent first historical novel, "Gods of War." Diana and I invited him to dinner at the Ritz and we had a lovely evening with his wife and him. I asked him how it was possible for him, who had not experienced the event to write so moving a description of the Bataan Death March — the best I had ever read. It really made me relive the affair. He told me that he had attended several meetings of the Survivors of Bataan and Corregidor and ten or fifteen years earlier, had made the march himself with a jeep following him and with plenty of water. But he experienced the brutality of the heat and sun and the climb up the hill from Mariveles to Cabcaben firsthand. I was impressed!

Jonathan Mark Goldblith as Cantor at a Wedding for the First Time

Steve Tannenbaum asked me whether I thought Jonathan would be willing to be the Cantor at Lisa's forthcoming wedding to John Hanselman. I suggested that Lisa speak with Jonathan directly. I could not speak for him. When she approached him he demurred saying that he had never done a wedding before, that he did not have the music nor know the ritual and stated: "Lisa, I've only been to one Jewish wedding before — my sister's" — and that was in 1969! Lisa persisted and, in the event, Jonathan *was* the Cantor at the Tannenbaum-Hanselman wedding at Temple Israel in Boston on October 12, 1985. He was magnificent and Diana and I thanked the Good Lord for blessing us with a son with such a magnificent voice and who was willing to share his talents with others. We were proud!

Visit of Future Crown Prince of Japan to M.I.T.

On October 21, 1985, Prince Hiro (Hiro-no-Miya), paid a scheduled visit to M.I.T. In the absence of President Gray, who was to be on a business trip, I was asked to plan the program, meet Prince Hiro and escort him. I met him at the 77 Massachusetts Avenue entrance and escorted him to several laboratories. David Gossard put on quite a demonstration with his graduate student assistants of CAD CAM. Later that evening, Diana and I were invited to dinner with Prince Hiro at the residence of Consul General Taniguchi which was a memorable affair. Former Ambassador and Mrs. Reichhauer were there, as well as other notables.

Some 1985 and 1986 Activities

At M.I.T., in addition to our normal business in Resource Development, we were beginning to plan for the *next* capital campaign, preliminary to which it was necessary to obtain an idea of priorities from the various deans and vice presidents. This was quite an undertaking, I knew then that I must step down soon, inasmuch as I was getting too old to undertake the rigors of another capital campaign and had no desire to. President Gray and I discussed the matter and he agreed.

In October 1985, I went on a rush trip to Japan via Bjuv, Sweden. I wanted to be in Bjuv to pay tribute to one of my oldest and dearest friends in Nestle, Mr. Karl-Evert Flink for whom a 70th Birthday dinner was being held. It was a lovely event. Then I flew on to Japan via Alaska. In all, I was gone but twelve days. Then Diana and I spent ten days over Christmas in Bermuda.

I was in Japan again in March, 1986 to help celebrate the Tenth Anniversary of the founding of the M.I.T. Japan Office. The I.L.P. put on a first rate symposium on “Emerging Technologies: Visions and Opportunities,” I spoke as did President Gray. In President Gray’s remarks, he mentioned that on June 30, 1986, I was stepping down as Vice President to become Senior Advisor to the President (for Japan) as well as continuing to be Professor of Food Science. To me, it had been a demanding job to be a senior officer at M.I.T. for eight years, something that I never dreamed of when I was a student or even ten years earlier — but most satisfying. My father would have been proud.

I made one more trip to Japan in 1986, in November, where I attended the 75th Anniversary of the Founding of the M.I.T. Club of Japan, which took place at the International House — a

FIG. 17.25. HON. TAKEZO SHIMODA OFFERING THE TRADITIONAL "KAMPEI" (TOAST) AT THE CELEBRATION OF THE 10TH ANNIVERSARY OF THE M.I.T. I.L.P. PROGRAM, MARCH 1986

FIG. 17.26. THE GRAYS AND THE GOLDBLITHS AT THE RECEPTION FOR THE 10TH ANNIVERSARY OF THE M.I.T. I.L.P. IN JAPAN

memorable occasion. I also had the pleasure, on this visit, of introducing Dr. Erik Von Sydow, of the Nestle Co., to the then Foreign Minister of Japan, Rt. Hon. Tadashi Kuranari, whom I had known well for over a decade. Both Diana and I were on friendly terms with both him and his wife and two boys — one of whom went to graduate school at M.I.T. and the other to Harvard. Diana and I saw the Kuranaries again a few months before Diana died.

CHAPTER 18

SENIOR ADVISOR TO THE PRESIDENT AND PROFESSOR OF FOOD SCIENCE (JULY 1986 TO JUNE 1990)

In July 1986, I moved from the Vice President's suite, Rm 4-202, to a more modest but wholly adequate smaller space in Room 10-282, which had just been freshly repainted, new carpet, desk, etc. It had problems with the ventilation but those were taken care of in due course.

Obvious changes occurred, which I felt would be better for me — less mail, many fewer phone calls, many fewer meetings, much less demand on my time — it was a quieter existence to which I would become adjusted. I had been on a treadmill since

FIG. 18.1. SAG AT DESK IN VICE PRESIDENT'S OFFICE CIRCA 1985-1986

1974 and although wearying, it was still a joy to get up in the morning and go to work each day.

I continued to prepare for, aided by George Sankey and Mrs. Saito, particularly, two trips per year to Japan. I also made two or so trips per year to Europe for the Nestle Company. Diana spent most of February, 1987 in Bermuda which afforded her great relief in breathing. She was to do this again the following year. The only bad part about this arrangement was loneliness in Bermuda, which I tried to alleviate by phone calls and spending a long weekend there. She stayed in a lovely small apartment on a side street near the direct bus line to Hamilton and a few minutes walk from a small superette. It was an ideal spot.

On April 3rd, there was a festive symposium in honor of my good friend George Buchi and the announcement of the establishment of a Buchi Lectureship — a most fitting tribute to this genius of an "organiker."

Sarah Beth Clark's Confirmation

On June 2, 1987, Sarah, our elder granddaughter was confirmed at Shavuoth (Pentecost), at Temple Tifereth Israel in Malden. She had done exceptionally well in her Hebrew studies and much credit must be due, in addition to her own initiative, to our Hebrew School and its teachers at the Temple and to Rabbi Stuart Pollack. She certainly is a scholar in the field now, at least from where I sit.

On June 3rd, Diana and I left for three weeks in Europe and while there we saw old friends in Zurich, Dr. Arthur and Deena Blumenthal, whom we had met years earlier through the Pilniks. Deena's parents, the Leo Orlows, were in the retail fur business, together with his brother, and had a store, Pelz AG, on the Bahnhofstrasse. Diana had purchased a fur coat there in the late 1970's and Leo Orlow had remade an old mink coat, that my father had given Diana after he had moved into our house to live with us, into a shorter sportier belted back jacket. It was beautiful.

FIG. 18.2. CONFIRMATION OF RACHEL DEBORAH CLARK
MAY 29, 1990
(L to R) SAG, Sarah, Diana, Rachel, Jonathan, Judy and Rich Clark

It was Leo who had recommended Flims Waldhaus to us as a wonderful place to eat and stay. I rented a car when we were in Switzerland in the late 1970's, for a week or so, and we went to Chur where the language spoken is mostly Romantsch. We explored the beginnings of the Rhein River, "Vorderrhein," not many miles from Flims. I am glad that I rented a small car for the narrow dirt roads. It was a pleasant trip.

We always enjoyed our visits with the Blumenthals very much and I continued to see them until I stopped traveling in 1993, because of difficulty I was beginning to develop in walking.

Bat Mitzvah of Rachel Deborah Clark

Now, the "baby" of our small family was to be Bat Mitzvah. On Friday evening, June 26, 1987, Rachel chanted the "maftir" and "Haftarah" for Parsha "Korach" at Temple Tifereth Israel in Malden. She did it beautifully and in full and ancient form. Judy and Rich provided a magnificent "Oneg" afterwards and Diana and I "qwelled."

Diana spent most of February, 1987, in Bermuda again, and I went to be with her for a week or so, and Rachel came down for her February vacation period. The stay seemed to do Diana a great deal of good in terms of her asthmatic condition.

Another Planning Session with George Sankey, August 1987

On August 31, Diana and I flew to Los Angeles to meet with the Sankeys and to plan for our next trip to Japan in the fall. While there, I attended my first meeting of The "Zentsujians" to which I belonged. As the name implied, it is an organization made up of "alumni" of the Zentsuji WWII POW camp on Shikoku Island. We had a grand time. The Sankeys drove us there and stayed with us. We saw the Leggetts and the Wolfsheimers and the Schwartzs among others. It also was my first trip to Disneyland. I should mention that George Sankey and I did do our work in L.A. planning for the November trip before we went to Anaheim.

High Voltage Engineering Corp. Takeover

In 1987, a hostile takeover bid was made by a group with funding from Beresford of England. This was pursued by them and resulted in endless board meetings through a part of 1988. I recall that on January 19-20, 1988, while in Vevey, Switzerland, I took part in a HVEC. Board meeting, via telephone, for four hours from 10:15 PM to 2:15 AM on the 19th and 20th (Swiss time) of January. Alas, all to no avail. The company was acquired, dismembered and is now no longer an entity. Very sad! On April 24, 1988 Dr. Pascal Levesque, CEO of HVEC had the Board members and their wives for a farewell dinner at the Pillar House Restaurant. It was nice but again very sad. He was a super executive and he had remade that company into a profitable entity. In the event, in 1988, the High Voltage Engineering Corp. was taken over in this unfriendly bid. The law just does not protect the right, necessarily. It is the dollar that will win out, in the end.

A Week in Interlaken, Switzerland

In June, while on a Nestle trip to Europe, we spent a week's vacation at Interlaken proceeding via the M.O.B. scenic mountain railway from Montreaux, over the mountain to Gstaad to Zweisimmen, and thence via another train to Spiez and Interlaken. We stayed at a smaller hotel, the Hotel Du Lac, where the rooms away from the RR Station were very nice and quiet, (we learned this after the first night and asked to have our room changed). We made a number of day excursions from the Du Lac, including a boat trip around the Lake of Spiez, a trip to the Freilichtmuseum Ballenberg, an open air museum containing a number of Swiss houses from various villages all over Switzerland — a remarkable and unusual museum, located near Brienz. We spent another day at Schynge Platte and Trummelbach Falls (where Sherlock Holmes met Prof. Moriarty), a day to the Jungfraujoch via a remarkable cog railway right into the mountain where the terminal is at 11,400 feet. On our last day, George Buchi had arranged for us to meet his brother Heinrich and his wife, and they took us together with our baggage to Speiz where his wife grew up. We saw the wood carving shop at Brienz (wood carving is an art form for which Brienz is famous). Then they drove us to a wonderful hotel where we had lunch and on to Bern where Diana and I took the train for Zurich. They are lovely people and we appreciated very much their kindnesses. It was a wonderful week's holiday.

Over the years, we have spent a number of one-week holidays in Switzerland in disparate places such as Braunwald, in Flims at the beautiful Flims Waldhaus Hotel, once in Ascona on Lake Maggiori one May where it rained five out of the seven days, at Interlaken and on two separate occasions at the Hotel Margna at Sils Baselgia (in fact we planned to go there again on the trip when Diana passed away in France). Each of the above places has its beauty and charm. We learned about Braunwald from Ray and Gay

Pariser. Ray's parents were staying there at the time and we had a great time with them on a couple of occasions.

We helped Prof. Sat Masamune celebrate his 60th (Kanreki) birthday — a big event in Japanese culture at a formal dinner at Maison Robert in August, 1988.

Our dear friend, Dr. Sid Levine passed away in 1988 in early September and I passed into the caring hands of Dr. Michael A. Kane of the M.I.T. Medical Department on September 15th, when I had my first appointment with him as my primary care physician. In this, I was truly blessed and am eternally grateful.

Max B. Greenberg Passes Away

Diana's Uncle Max, her father's youngest brother passed away on November 27, 1988 after a long illness brought on by his hip giving out, while he was repairing his picket fence. He lived a long life, loved by us all. But he was lonely. Diana and her cousin Donald Van Greenby were the ones who visited him the most and Diana was super and meticulous in visiting Uncle Max when she visited her mother. I joined her on the Saturday afternoon visits. He was a great guy, a veteran of the World War I Balloon Corps, a former actor in Lowell, a featherweight boxer at age 15, always smoked cigars, had been a garage owner and sold Hupmobiles. A proper biography of him would really make interesting reading.

Jonathan Mark Goldblith as Tenor Soloist and Cantor

Earlier I had referred to Jonny's vocal capabilities as a tenor. He had been in the choir of Temple Tifereth Israel of Malden since about age fifteen. On the advice of Mr. Wood, the music teacher at the Melrose High School, he took voice training from a Mr. Frank Terminello, who was recommended to us, and Jonny began to develop as the top tenor soloist of the choir.

In 1987-1988, Cantor Chandler at our Temple took a well-earned sabbatical, and Jonathan substituted for most of the Friday evenings of the year between Succoth and Shavuoth. Diana and I attended almost all of these and I taped (audio tape) all that I attended. He had developed into quite a tenor cantor. In the meantime, he had served as cantor at two weddings in our temple (in addition to his first wedding performance at the Tannenbaum-Hanselman wedding in Temple Israel).

For the past two years (1994, 1995) he has had the role of the mainstay of the cantorial group at the High Holiday services — a trying job especially while living in Maryland for the past several years. Part of his success is due in no small measure to his friend (and ours) Mrs. Bryna (Bunny) Tabasky, our choir director, who has sent him music, tapes, etc., and helped with extra rehearsals whenever possible.

Bunny and Jonny did two Bat Mitzvahs for the Tabasky nieces in a Temple in the Florida Keys (niece of Bunny and Sam Tabasky) over the past several years. The "balabatim" there are ready to hire him for the High Holydays! His voice has gotten even better now that he sings with the choir of the Episcopal Church in Frederick, MD and vocalizes every week. As Diana once told him, the important thing to remember is that God gave you that beautiful voice. Now, it is up to you as to what you do with that talent. Jonny has given many of us, especially Diana and me, much pleasure therefrom.

Fiftieth Anniversary of The Founding of the Institute of Food Technologists

In about mid-1987, I was approached by the Institute of Food Technologists to write about the history of food science and technology and the developments therein over the past fifty years. It would be presented at the Annual Meeting of the Institute scheduled for June 1989, in Chicago. I agreed to do this and needed all the time between being asked and the time it was due. In the event, an eleven page article, printed two columns wide, was published in *Food Technology* the following September 1989. I presented orally a slightly different version. The paper was published as: "50 Years of Progress in Food Science and Technology: From Art Based on Experience to Technology Based on Science." This was one of the few plenary event papers and I was again among old friends and colleagues. I was thrilled to be asked and am glad that Diana, who had been with me to so many earlier I.F.T. affairs and meetings, could be there to hear me. She was my severest critic and staunchest defender, always.

A Word About the M.I.T. Club of Japan

The M.I.T. Club of Japan is a remarkable organization. All who went to M.I.T. in some capacity, either as students, post-docs, laboratory workers, etc., are eligible for membership. It is, I believe, the oldest M.I.T. alumni organization outside of the U.S.A. but not limited to former students only. When I began going to Japan regularly in 1976, they would often hold a meeting around the time that I was to be there, or President Wiesner, or President Gray, or Chairman Johnson. The meetings were usually at the International Club or the American Club, and when I happened to be in Japan at that particular time, I attended. In the 1970's and until his death, Dr. Ayukawa was the unofficial advisor to the club. During "my watch," a Prof. Hori was the president and later and during the 75th Anniversary, Prof. Hiroya Fujisaki of the Dept. of Electronic Engineering of the University of Tokyo was the president. He was followed by Dr. Hajime Mitarai, President of Canon Inc. who served until his sudden and untimely death in the early 1990's.

The 75th Anniversary Book published by the club has a number of early, as well as more recent pictures therein. One of the most well-known members was Rear Admiral, (Ret'd), Yoshio Kubota, whom I got to know pretty well in the 1980's. Admiral Kubota was an alumnus with an S.B. and an S.M. in Mechanical Engineering who became Fleet Admiral Yamamoto's Chief Engineering Officer of the Combined Fleet during World War II.

A number of faculty of the Institute and officers have spoken at the meetings of this club — even before airplane travel overseas. In 1933, when Prof. Joseph S. Newell, of the Dept. of Aeronautical Engineering and then an expert on nosewheel design, spoke, one of the attendees was Admiral Isoroku Yamamoto. In 1934, Prof. Norbert Weiner spoke with Mrs. Weiner present also. Prof. Weiner spoke again in 1956. Other speakers post-WWII included Dean Hazen and Prof. Wiesner, in 1963 when he was Science Advisor to the President of the United States. Dr. Killian and Dr. Zacharias spoke in 1967. General Lampert spoke in 1973, when he first started to come to Japan on behalf of the Institute. In 1986, I represented the Institute at the 75th anniversary dinner of its founding. It was an illuminating experience. One could not ask for a more loyal group of alumni and former associates of M.I.T.

Chair Established in SAG's Name, November, 1989

In November 1989, the announcement was made by President Gray, in Tokyo at the All Nippon Airways Hotel, while we were there on a business trip, of the establishment of an endowed chair in Applied Biology in my name. I was astounded and Diana was thrilled, as was I. This was all done through the hard work and initiative, I learned later, of two good friends, Dr. Yaichi Ayukawa and Fred Gross. Each solicited on their side of the Pacific. They presented Diana and me with a sterling silver salver, suitably engraved, and I was able to thank most of the Japanese donors who were present at the coffee/cake ceremony. I was so happy that Diana was there also, as was Priscilla Gray and Masako Ayukawa. To Yaichi and Fred, I owe an irreparable debt of gratitude for all they did to bring this about — no easy task at all as I well know.

Diana's Last Visit to Japan, March 1990

In January 1990, we both went to Paradise Island to the ILSI-Nutrition Foundation meeting held in Paradise Island, Nassau. It was a good five day trip away from the cold climes. In

March 1990, Diana made her last visit to Japan with me. The two highlights of the visit, in addition to seeing, of course, our dear friends the Ayukawas, Mikis and Takasakis, included first of all a visit to the Togu Palace to call upon Naruhito, as he was now called, the Crown Prince, formerly Prince Hiro-no-miya, whom we had last seen at the home of the Consul General in Boston on October 21, 1985. It was good to see him again and we spent a pleasant half hour or forty five minutes with him. He presented me with an inscribed copy of his bound Research work on the Thames as Highway (in the Eighteenth Century) inscribed to me by "Hiro," which I value very much.

FIG. 18.3. DIANA L. GOLDBLITH AT ODAWARA, JAPAN DURING VISIT WITH MITARAIS — HER LAST VISIT TO JAPAN, MARCH 18, 1990
(Courtesy of Mrs. Hajime Mitarai)

The second was a visit to the country home of Michiko and Hajime Mitarai in Odawara on March 18, 1990. It is a lovely place with a carp pond, gardens, a beautiful old home that Hajime inherited from his father, Dr. Tadashi Mitarai, the founder of Canon Inc. While on this visit, we were taken once again for lunch at a lovely small French-type bistro in Odawara that Hajime's father had discovered. There, Michiko took the picture with date thereon that appears in this book. It is the best I have of Diana in her later years, is in color, and Michiko thoughtfully had several copies enlarged for me. We had a wonderful visit with them on a beautiful sunny day. This was to be my last official trip to Japan as an employee of M.I.T., although not my last visit to Japan.

FIG. 18.4. LAST PICTURE OF DIANA AND SAG AT M.I.T. GRADUATION CLASS OF 1990, SAG'S 50TH REUNION, JUNE 4, 1990
Taken at Killian Court, M.I.T.

Confirmation of Rachel Deborah Clark

On Shavuoth, May 29, 1990, Rachie was confirmed at Temple Tifereth Israel in Malden at a lovely ceremony. This would mark the last of the religious ceremonies for the two "ladies," as I used to refer to them when they were little girls, until the marriage ceremony. It was a lovely evening and Rachel did us all proud (as usual with both Sarah and Rachel).

Death of Diana L. Goldblith

On June 8, 1990, Diana and I left for a trip combining business and a bit of pleasure. The business portion was to be in two parts before and after a week at Sils Baselgia at the Hotel Margna. We arrived in Paris on the morning of June 9, 1990 on the overnight TWA flight. When we arrived at the Hotel Lotti, where we were to stay for several nights, Diana partially undressed and said that her asthma was bothering her and she would lie down for awhile. I stated that I would reconfirm our onward reservations across the street at the air booking office. I returned within thirty minutes or less and she told me she was in trouble. I asked her if she had taken some prednisone and got up to get some in the bathroom, and when I looked over my shoulder she was gone. I yelled for help and began emergency heart CPR. The French EMT's must have arrived within ten minutes and continued to work on her for over an hour. But to no avail. She was gone. I was bereft. I called Judy and Rich. The phone calls went back and forth when I was in the room.

Meanwhile, I had to make a statement to the police and then had to go the headquarters of the district to make another statement which was typed and I signed it. I was in a daze. Judy called me back and said that Jonny would be over the next morning on the same plane that we had taken. I tried to tell her that there was no need to but Judy, Jonny and Rich insisted. In the end, I was so happy to see him.

As Jonny was getting through the line to Immigration in Paris to get his passport stamped, he ran into Judy Lindahl who was also arriving for a two week's vacation in Paris. Jonny told Judy why he was there looking so glum. She said, "Tell your father I'll be over later. He'll probably need some help from a lawyer and I loved your mother (or words to that effect)."

I will never be able to repay Judy for her many kindnesses those next few days. She dealt with the U.S. Embassy and the myriad of details attendant to shipping remains overseas. Jonny and I just dealt with the Parisian undertaker. Nestle provided the up-front cash needed (I had cash but not in that quantity). Nestle sent their head of the Legal Department in Paris. He was very helpful. Judy Lindahl was present at the closing of the casket and had a Rabbi present and appropriate prayers said. Thus I could go home on Monday or Tuesday. What a tragic end for poor Diana.

On Sunday, June 17th, 1990, she was buried after a Funeral Service at the Temple. Rabbi Stuart Pollack gave a magnificent eulogy. Our friends pitched in at our home to help after the funeral — young friends from M.I.T. and older ones from there and from Melrose. Older friends of Diana and mine. All came to the house for Shiva. She was loved by all and had done so much good on this earth during her pilgrimage here. I miss her so very much as do our children and grandchildren. May God rest her soul.

Mike and Lisa Kane and Milton and Ruth Issenberg had me over for meals. Everyone was super to me. But, the void was there and is still there as I write this page, in May 1996.

CHAPTER 19

RETIREMENT

Move to Building E-38 from Bldg. 10-282

In late 1990 or early 1991, I was asked if I minded moving over to Building E38 where the I.L.P. was located. I was given a choice of several offices and selected the corner office on the fifth floor. This move would enable Fred Gross, who was then Director of Corporate Development, to move his research team into the space I was occupying in Building 10-282 and would be close to his office. I readily agreed to this and my good friend Bill Dickson spoke to another friend, Gene Harmon, Director of Physical Plant, and they refurbished the 5th floor corner office for me overnight. It really looked good. My new I.L.P. neighbors were impressed at the quick service. Eric Johnson placed in the office the large curved desk which I had used as Director of Industrial Liaison. This desk originally had been part of the central reading room in the old main Library on the fifth floor in Building Ten.

Association with Cathedral High School

I hope during my retirement years from M.I.T. that I have been of "aid and comfort" to some of the I.L.P. staff as partial payment for the use of the real estate. Actually, after Diana passed away, I had no idea what I would do with my time. I knew that I had to keep busy at least five days per week as long as my health would permit and keep active professionally. Friends and circumstances came to the rescue. My physician and good friend and his wife, Mike and Lisa Kane, acquainted me with Cathedral High School in the South End of Boston. I was appointed a Member of the Board along with Mike Kane and we spent a few evenings there each year, trying to raise funding for this worthy cause to educate mainly minority students.

In the course of this activity, I came to know, admire and appreciate the Principal, Sr. Patricia Keaveney, SND, who was a remarkable person. She had spent part of the 1950's-1960's at St. Mary's Parochial School in Lawrence, and was the principal thereof near the end of her stay in Lawrence. So we had much in common. Sr. Pat was a lovely person, dynamic, vivacious, ready to take on all comers — even the Cardinal, for Cathedral High School, its welfare and that of the students. She had been ill for years, suffered, but worked at quite a pace until illness forced her to retire. Cardinal Law was a true admirer of her and a friend, and visited her in the hospital near the end, as well as celebrating her funeral mass. His eulogy really captured the essence of the lady. A remarkable person that I was fortunate enough to get to know. CHS was a project I enjoyed doing.

International Life Sciences Institute-Nutrition Foundation

I had enjoyed being a trustee of the Nutrition Foundation since about 1970, and continued on even after it was conjoined with the International Life Sciences Institute (ILSI). The spark plug behind ILSI, and its "father," was a former student in the department, Dr. Alex Malaspina, Senior Vice President of the Coca Cola Co. in Atlanta. The offices of ILSI are in Washington, DC with branches now almost all over the world. He has done a fabulous job. He wanted me earlier on to assist him in fund-raising in Japan, but I informed him that I could not do so while I was still working at M.I.T. There would be a direct conflict of interest. When my retirement drew near, he approached me again. This time I agreed and made several trips to Japan with him to introduce him to some people I felt he should know. Yaichi Ayukawa also helped him immensely in the last year or so of Yaichi's life.

In March, 1990, I got help from Alex for the M.I.T. Rugby Team, which was planning a visit to Japan. I was approached by Jim Culliton, the team's coach-player to help them with meals, funding, etc. I contacted a former student, Yutaka Saito, and through his connections we got them banquets at the Lion Beer Halls. Alex connected me with one of the Japan Coca Cola Co.'s vice presidents, Dr. Tadashi Togami, whom I knew already and they set up a first rate “bash” for the ruggers at the Tokyo American Club, where I happened to be along with Diana and the Ayukawas at a small farewell party the Japan Office had for me. It was a joyous occasion and the Rugby Club presented me with a sweater and jacket with the Club/MIT logo thereon and “No. 1” on the sleeve. I have enjoyed wearing it to Temple occasionally on a Sunday morning. Some must think that I was quite an athlete in my youth!

Chance Meeting with HIH Hiro-No-Miya

One other point regarding Alex and one of our trips to Japan relates to the time in March 1992, when Alex and I were at the Okura and we saw a red carpet at the front entrance. I asked Mr. Goro Yamazaki, the General Manager, whom I knew very well, what the reason was for the red carpet. He informed me that the Emperor (Akihito) and the Empress were attending a reception downstairs. I had met the Emperor and Empress a few years earlier, when they were in Boston when he was still the Crown Prince. So I suggested to Alex that we “stick around” and see who else was coming. All the guests arrived first and seemed to be the leading lights of the Japanese Industrial World. I spoke with Dr. Soichiro Toyoda, President of Toyota Motor Co., Mr. Ishihara, President of Nissan, Mr. Kaneda, President of Nichirei, and a few others. Alex was impressed. Then came HIH the Emperor, and the Empress, and then their eldest son, Prince Hiro. Instinctively I exclaimed “Hiro no Miya”! He looked up smiled and went on with the family. About five minutes later I was approached by Mr. Yamazaki informing me that HIH Naruhito asked if I would be kind enough to come see him for a few minutes downstairs.

I suggested to Alex that he come with me. He demurred at first saying that he was not dressed (the guests were all in dinner jackets and long gowns), but in the event Alex came along and met the Crown Prince, who was most gracious and cordial and we saw his parents as well. Alex has never forgotten this chance meeting with the Imperial Family of Japan. I made several trips with Alex to Japan. We concluded the consulting relationship in mid-1994, I believe. It was fun and I hope useful to ILSI. It also kept me going to Japan on worthwhile purposes. Alex and his people ran some symposia in Tokyo and Nara as well. He has done remarkably well in building up ILSI as a potent and useful international force to achieve harmonization of food and food additive legislation. I am proud of his accomplishments.

Visit of Atty. Odette Dupont of Nestle

In January 1991, Atty. Odette Dupont who had assisted Mr. Ferdinand Elsener in the Legal Dept. of Nestle and had assumed his responsibilities for the United States as the chief legal advisor, visited me for a few days. We did a small tour of Boston via car, trolley bus (the first time I had used the latter), which was a good way to see the city sights. We had dinner on Saturday night and had Judy, Rich and Judy Lindahl. Odette had been most helpful to me when Diana passed away in Paris and lined up all the help for me, made the cancellations, etc. Both Attys. Lindahl and Dupont were wonderful to me as were Judy Clark, and Jonny and Rich. Judy Clark took care of all of the estate problems for me in a record and timely fashion, and with a professional exactitude and flair which made me very proud. I took Odette to Portsmouth, NH for a small tour around that area and some more good seafood before taking her to the airport on January 17th, 1991.

Miscellaneous Items, 1991

In the meantime, various friends were having me to dinner and were making sure that I was kept busy. As I have stated earlier, tops outside of the immediate families were Ruthie and Milt Issenberg and Lisa and Mike Kane. My constant luncheon companions were either George Buchi, Fred Gross, occasionally Bill Dickson, Susan Crowley, Karen Tenney, and a host of others fairly frequently. Buchi and I would go to Locke Obers or Maison Robert in 1990-1994, and when there was an article in the Globe about regulars at Locke Obers, George's and my names appeared at two tables. My friends (and I also) got quite a kick out of this.

At the Passover Seder on March 29th, our first without Diana, we asked the Kanes to come and they have come almost every year since. They really got into the swing of the melodies. Occasionally, in 1991, I drove to Bridgeport to visit with Diana's cousin, Gail Gartel, who had been very kind to us over the years and very close to Diana as she was growing up. I took Gail and her friends to lunch. I felt good doing it and enjoyed myself as well.

Rachel was inducted into the National Honor Society on May 9th and I went along with her proud parents and "qwelled" as she walked away with many citations. Sarah was already at Wesleyan and very busy as a student in the College of Letters.

In May 1991, I visited Europe again for Nestle and attended the opening of the new "reco," (Yorkreco) at York at the Roundtree subsidiary company, where most of the chocolate and cacao bean research and development was to be done. I also had an opportunity to see the manufacture of the "Kit Kat" bars. What a mammoth operation and complicated as well. But oh, so ingenious and efficient.

While I was in London, Irene and Donald Walker invited me to Stanmore to their new home on the second day of Shavuoth on May 20, 1991. I was also able to go to schul with Donald in the morning so that I could say Yiskor. They had some friends over for kiddush after services. It was a most pleasant day.

In London, on that trip, I had a reunion with Dolly and Gary Rosen. Poor Dolly, she was ill and died not too long thereafter. She was a great person and a superb hostess. I am glad that I saw her once again. From London, I flew to Copenhagen and Bjuv, commemorating Diana's first Yahrzeit at the Synagogue at Copenhagen. Jim Flink, a former student and colleague, who knew Diana well, came with me. I think it was the first time Jimmy had attended a "mincha/Mariv" service carried out so fast that they were over before he realized what was going on. But there was a minyan and I got to say Kaddish. The next day, May 29, 1991, after the morning service, I took the plane to Paris and met Dave Davis with whom I went to Liseux and to Tours, to the recos at each of these laboratories there. I then proceeded home, on June 2nd, to Boston via Zurich.

In the latter part of June, I was able to have Mr. Karim Abdullah of La Tour de Peilz, where he has a pharmacy, for a visit to Boston. He has been very good to me over the years, since I have been going to Vevey, and formerly worked in the Pharmacie de Trois Couronnes, across the street from the Hotel des Trois Couronnes where Nestle always put me/us up. Karim has been most helpful to me when I needed some medical aid or ran out of pills, etc. He was here to bring his teenage son to the Taft School in Connecticut for the summer for a special program they have there.

Joining Congregation Tifereth Israel (formerly Tifereth Anshei Sfard)

During 1991, following the “Aufruf” of Dick Issenberg before he married Peggy Lenz, I attended several Saturday morning services at my parents’ former schul, now the Congregation Tifereth Israel. I enjoyed the services and especially being with Milton and some old acquaintances. Most especially, I enjoyed the Rabbi, Barbara Penzner, who was a graduate and received “smicha” from the Reconstruction Seminary founded by that innovative 20th century thinker and scholar, Rabbi Mordechai Kaplan. Rabbi Penzner was outstanding in every aspect of her ministry in Lawrence/Andover — sermons, knowledge of talmud, torah reading, davening, etc.

So I joined this congregation in 1991, and have gone to Saturday morning services there on a more or less regular basis since. The Congregation sold their building in Lawrence inasmuch as most of the membership was now from Andover anyway (with only a few from Methuen and Lawrence). First they rented space at a church on Central St. in Andover and last year purchased a home with a goodly amount of land, reconstructed the home so that there is a sanctuary that will seat about eighty people there. Plans are underway for a further enlargement at the present time (May, 1996). It is once again a young dynamic schul, which is Conservative, and a lovely group gathers for services every Saturday morning. I see an old friend who knew my parents well, Lillian (Rosenberg) Shlossberg, almost every morning that I attend services. She is now over ninety years young. I do the Maftir whenever I am asked to and it is a pleasure to listen to some of the members lead the davening and/or chant the maftir or Torah reading. And, of course, it is a pleasure to be with Milton, who has been going there for years.

The Legacy of Columbus with Particular Reference to Foods

In 1991, early in the year, I was asked by Neil Mermelstein, the Senior Associate Editor of the journal *Food Technology*, if I would care to write an article to be published in October, 1992, celebrating the five hundredth anniversary of Columbus' original voyage across the Atlantic Ocean. I agreed to do it. It took me over a year to research and write. I found most libraries devoid of the material I would need. I haunted the local Boston/Cambridge bookstores and in the end purchased over fifty books published since 1990, on Columbus and his voyages. No wonder the libraries did not have them. I wrote, researched, rewrote and finally ended with an article of thirteen, double columns, printed pages, superbly illustrated by Neil Mermelstein with the U.S. Postal Service stamps issued in May, 1992. The article was featured on the front cover of the October 1992 issue of *Food Technology*. I will present here only the cover subtitle: "Columbus's voyages to the New World made possible a gastronomic treasury and a nutritional status never before dreamed of." I felt pretty good about my main writing accomplishment of the year, 1991-1992.

Samuel Cate Prescott: MIT Dean and Pioneer Food Technologist

Having written and published this opus on Columbus, I felt that I should keep occupied and writing was something that would certainly occupy the mind and could keep me as busy as I cared to be. The truth of the matter is that once I get started on a writing project, I cannot wait to get it done. In the event, I began to collect material at the M.I.T. Library Archives on Dean Prescott. I had long admired the dean and felt that if I did not write something in detail about his life and work, the likelihood of someone else doing it was not very great. Moreover, I had written a long obituary of him for the journal *Food Technology* and had an idea of the amount of material there might be.

Accordingly, I began to do some research preparatory to writing a biography of Samuel C. Prescott, M.I.T.'s first Dean of Science and considered to be one of the early pioneers of modern food technology. My purpose here is not to outline the book but rather, in a short space, to summarize how I went about doing it. I went to the M.I.T. Museum off Massachusetts Avenue and went through the material they had on Prescott and his teachers and colleagues. Then I went through, as one finding led to another, to the M.I.T. Archives material on Prescott, Compton, Killian, and relevant faculty colleagues of Prescott's. I continued to plow on both in terms of writing and in more searching. I called the former Sanborn Seminary in Kingston, NH and this led to a prominent alumnus who knew where the old Prescott house and his birthplace was. I went to South Hampton, took pictures, etc., located Prescott's parents final resting place, and in the process learned that the local church in South Hampton was not aware of his parents' resting place in their churchyard.

I located his daughter, Eleanor, in Johnson, Rhode Island and she kindly let me look at all material she had about the Prescotts which was very helpful and much appreciated. I began to write early in 1992 and would get just so far and then back to more research to answer a particular question. I imagine that I wrote fairly steadily for five or six months (note: by writing I include handwriting, and mainly on two or three computers that I had).

Probably the catalyst that got me really started in earnest on the Prescott bio was a telephone call from Guy (Dr. Gideon) Livingston, who was planning a "Pioneers in Food Science Series" and asked me if I would do one on Prescott and Proctor. I told him that I would do Prescott alone, but that I knew what there was on Proctor and it just was not enough material to do a complete booklength biography, and I did not know where to get any more material. Then I made up my mind that if there were to be a bio of Dean Prescott, it should be the first in the series and I worked intensely at the Prescott bio, for awhile. I contacted the publisher, Mr. John J. O'Neil, Publisher of the Food and Nutrition Press, and he and his associate cooperated beautifully and fully.

Mike Kane edited the whole book before it went to the publisher for final editing. He was great and I don't know where he found the time to do this as well as his other obligations. He's a wonderful friend, as well as an excellent editor!

The book was published in May or June, 1993. I don't think it was a best seller, to say the least, and I was the best customer, giving copies to many friends and colleagues. I only wish that there were sufficient material for a biography of Bernie Proctor. Reviews of the Prescott bio have appeared and the major ones in *Food Technology* and *Technology Review* were fair and on target for the specific audiences for those magazine/journals.

Death of Dr. Yaichi Ayukawa

My oldest and dearest Japanese friend, Dr. Yaichi Ayukawa, passed away on November 30, 1991 after a short illness. I had seen him earlier in the year and he had complained of dizzy spells. However, he suddenly lost consciousness one day and soon was gone. What a tragedy for his dear wife, Masako and family, as well as those of us who were privileged to have known him over the years. I knew him as a student, a colleague and a dear and close friend for over thirty years. We vacationed together as families and celebrated good and suffered bad together. I lost a dear and close "tomadachi."

In January 1992, I attended his TechnoVenture Company organized funeral, representing the Institute, and read a personal message as well as the official Resolutions of the Corporation of M.I.T. on his death. It was a moving ceremony.

Sarah came with me on this visit to Japan in January 1992. We had a splendid time, although the initial purpose of the visit was a sad one. My good friends helped to entertain her. Mrs. Hajime Mitarai (Michiko) ran a dinner party at her home for Sarah and me, and, their daughter, Yumiko (Yumi) was home for the intercession from Gaucher together with Hajime's niece of the same age. The girls had an evening of fun. Their good friend and physician, Dr. Jiro Wada, a renowned heart surgeon and pioneer in open heart surgery, who had worked with Dr. Dwight Harkin in Boston shortly after World War II, was there together with his lovely wife and we all had a grand time. One day, I arranged, through Mrs. Saito, for Sarah to go to Kyoto on the Shinkansen. She was met in Kyoto by a friend of Mrs. Saito's and shown around, much more knowledgeably than I could have done it.

My friends the Takasakis and Mikis had a dinner for us at Pruniers and Sarah got to meet many of her grandparents' close Japanese friends. While the trip to Tokyo was only for one week, Sarah got to see quite a bit and meet a number of nice people.

Last Visit to Europe as Nestle Consultant, March 1992

My last official visit to Europe as a consultant of the Nestle Co., for whom I had worked in various capacities for almost thirty years, took place during the first seventeen days of March, 1992. I began in Copenhagen/Bjuv, and then went to Kempthal, and then to the Vevey area to visit again the Linor, Konolfingen, and the new Research Laboratory at Vers Chez Le Blanc. During this trip I had an opportunity to visit with Odette Dupont at her flat in the Valais near Sion.

The highlight of the trip was a small private dinner held on March 10, 1992 at "Hotellerie de Chatonneyre in Corseaux" near the former home of Charlie Chaplin. It was hosted by Brian and Maria Suter and included Carl L. Angst, Birgitta and Karl-Evert Flinck, Mary and Andy Ledzion — a select group of my closest friends of longest standing in Nestle. Brian presented me with a Nestle "retirement" mantelpiece clock which is beautiful. It is a Jaeger LeCoultre, shiny brass, modernistic in appearance, elegant and simple, with a small "Nestle" printed on the base, unobtrusively. The wine and speeches flowed and from me, probably a few tears. I had learned much through my association with this wonderful company and from its people. I made many friends and still hear from them and get invited to affairs held at the laboratories at Marysville and New Milford. This coming September, 1996, there will be another big affair celebrating the new laboratories at Marysville, Ohio and I have been invited by Elaine Wedral and the group. I hope I will be able to attend.

Later in that summer of 1992, Dr. Elaine Wedral, Chairman of Westreco/Nestle had a dinner for me in Danbury, Connecticut, and invited Jonny, Judy and Rich. It was a lovely affair — again many speeches. Dr. Wedral presented me with an unusual marble jewelry chest with a lovely inscribed brass plate in the cover. It was another memorable Nestle event. I only regret that Diana had not lived to have shared this with me. All present knew her well and all were fond of her, as she was of them.

First Occupant of Goldblith Chair Named

On September 24, 1993, the first occupant of the Goldblith Career Development Chair was named by the M.I.T. Provost, Dr. Mark Wrighton, and a celebratory luncheon was held and attended by a number of colleagues at the Institute and by my family — Judy, Jonny, Rich, and Rachel. Paul and Priscilla Gray were there and Paul and Mark Wrighton spoke. There were 60 people present and the climax was the presence of Masako Ayukawa and Junta Ayukawa, Yaichi's widow and son. It was a delightful affair. Dr.

Peter Dedon, a toxicologist, was named as the first occupant and I am sure has done the chair honor by his work and his character. Some nice things were said by Paul Gray about me which were greatly appreciated by me and my family. The credit for the chair goes so much to Yaichi Ayukawa and Fred Gross, as mentioned earlier. I am truly blessed.

FIG. 19.1. AT LUNCHEON ANNOUNCEMENT OF CHAIR IN HONOR OF SAG AT M.I.T., SEPTEMBER 24, 1993
L to R, Rachel, Rich and Judy Clark, Jonathan Goldblith and SAG, (Sarah was in Israel at the time).

Death of Dr. Hajime Mitarai

I learned from a telephone call from Michicko Mitarai of the sudden death of her husband, Dr. Hajime Mitarai, on August 31, 1995. He had suffered a bad cold and things went downhill from that. Poor Michiko and Hajime — they were such a devoted couple and lovely people who did so much for others. Diana was very close to Michiko, and Michiko introduced her (and later me) to the College Womens' Association of Japan, which was made up of women who went to colleges in the U.S.A., and whose purpose was to raise scholarship funds for poor students. They used to run art shows and seminar series, etc. Michiko was president one year in the-mid 80's and Hajime was so proud of her.

Hajime had been elected a Member of the M.I.T. Corporation in June, 1995, and both he and Michiko were looking forward to a trip to his first corporation meeting in October. Alas, it was not to be. They were so kind to Diana and me over the years. Now Yumiko, their daughter, will be enrolled in the Sloan School beginning September, 1996. Hajime would have been so proud.

Seventy-Fifth Birthday of S.A.G.

On May 5, 1994, I turned 75 years of age. Mike Kane and Steve Tannenbaum ran a luncheon for me at the Maison Robert and invited some of my closest friends at the Institute. That evening, Judy, Rich and Jonny had a dinner for me at the Boston Harbor Hotel, to which they invited Judy Lindahl (who shares the same birthdate). Both were lovely affairs which I retain fondly in my memory. Again, if only Diana were here to have shared the day with me. But such is life! I was blessed to have her for over forty-nine years. That is more than most people are fortunate enough to have. And a good marriage it was!

"Of Microbes and Molecules"

In 1993, when I published the Biography of Dean Prescott, I had an idea of another book. I felt that a History of Food Technology at M.I.T. should be written. I had, over the years, written a number of disparate articles containing snippets of the subject, but nothing that could be considered of any value or even much help — even were I able to locate them. So, back to the M.I.T. Library and Archives. I had reviewed the material I had collected on Dean Prescott and his own book, *When M.I.T. was Boston Tech.* I felt that to get a comprehension of the subject, I needed, first a brief synopsis of M.I.T.'s founding and early history. So, I went to both the M.I.T. Archives and the M.I.T. Museum to see what they had on William Barton Rogers and his family. Thus the first chapter was written.

Secondly, I needed to get a better handle on Prescott when he was a student. Who were his teachers? I knew who his students were and knew many of them. I recall Prescott's kind words about Ellen Henrietta (Swallow) Richards, who was one of Prescott's two most influential teachers (the other being Professor William Thompson Sedgwick). I began to read all I could in the Richards folders — both of Mrs. Richards and her husband, who founded the department that eventually became the Department of Materials. After writing the first chapter on Mrs. Richards, I became so immersed therein and so impressed with her accomplishments that I went on for three chapters about her and her researches — mainly the food and water work that she pioneered. She really began food technology at M.I.T. Actually, I considered writing a comprehensive biography of this remarkable lady, inasmuch as no good biography really exists. But I remembered my original purpose and stuck to it because if I did not do it, some years later the history of the development of food technology at M.I.T. would be done by some History of Science student or staff member, if at all. I was fortunate enough to have lived through many years of the era covered in the book. I resisted the impulse to write a bio on Mrs. Richards, which was not momentary, inasmuch as I was in correspondence with both Vassar and Smith Colleges, where much material on Mrs. Richards resides.

Getting back on track, once again, I then went into the Sedgwick Era, the Prescott Era, the Proctor Era, etc. The book was published in July of 1995. The only part of the story of the book that might be told here that is not self-evident relates to the cover. My friend, Connie Lewis, and my daughter, Judy Clark, told me about the mural on the side of the building on Newbury Street, corner of Dartmouth, featuring prominent Bostonians over the centuries. The important aspect to me is that of all the prominent people at M.I.T. since its founding, only Ellen H. Richards is painted in the mural — and she was never even a member of the Faculty, just of the Instruction Staff. I went over during 1994-1995 and took a number of pictures of the wall mural and closeups of Ellen H. Richards and those seated/standing around her. That was the basis of the attractive cover of the book.

The saddest part about which to write was that one chapter at the end where the Department of Applied Biology was closed down. It was not difficult for me to write about it from a composition point of view, inasmuch as I had lived through it and had the reports of the Faculty Committees investigating it. I used these reports as the basis of the chapter. It was difficult, however, on an emotional basis. But, we got through, and in July, 1995 it was published by John O'Neil, who did an excellent job. I had to pay for the printing, inasmuch as the expected funding from ILSI and the Sloan Foundation did not develop for differing reasons. I am still awaiting a major review in *Technology Review*, which as of May 1996 has not yet appeared.

In the meantime, I occupy myself as an observer and basking in the glory of my children's and grandchildren's accomplishments. Sarah graduated from Wesleyan University in Connecticut in 1993 and Rachel will graduate Cum Laude on June 2, 1996 from Williams College with an A.B. with honors in Studio Art and Psychology — probably to better understand her Grandfather's motivations and what makes him tick. Judy and Rich are both successful in their careers as is Jonny. I am proud of all of them and of their accomplishments. I still go into the Institute and will do so more frequently, D.V., now that these reminiscences are completed. I had been going in five days per week but the material for this present work is all at home. Now that I have just the tidying up to do, I will try to resume my former schedule (which I kept to until last January).

I get up each day, count my blessings that I am still able at seventy-seven years of age to still do so, a bit more frail, a little less steady afoot, but with God's help and Mike Kane's able medical care, I am able to get about, to still drive to Temple for an evening service when it is still during daylight hours and to Andover on most Saturdays and do the Maftir. I still read avidly. Truly, I can say a "Shechayanu." I am blessed. Will I write any more books? I don't know. I'll have to see if and when a writing urge appears again. At the moment, I am fully occupied finishing this volume.

Meanwhile, I take each day at a time, and pray that the Good Lord will grant me a bit more time to enjoy my children and grandchildren and friends. I hope that I will have left my "camping ground" in a little better condition than I found it.